PRESSURE GAUGE HANDBOOK

MECHANICAL ENGINEERING

A Series of Textbooks and Reference Books

EDITORS

L. L. FAULKNER
Department of Mechanical Engineering
The Ohio State University
Columbus, Ohio

S. B. MENKES
Department of Mechanical Engineering
The City College of the
City University of New York
New York, New York

1. Spring Designer's Handbook, *by Harold Carlson*
2. Computer-Aided Graphics and Design, *by Daniel L. Ryan*
3. Lubrication Fundamentals, *by J. George Wills*
4. Solar Engineering for Domestic Buildings, *by William A. Himmelman*
5. Applied Engineering Mechanics: Statics and Dynamics, *by G. Boothroyd and C. Poli*
6. Centrifugal Pump Clinic, *by Igor J. Karassik*
7. Computer-Aided Kinetics for Machine Design, *by Daniel L. Ryan*
8. Plastics Products Design Handbook, Part A: Materials and Components; Part B: Processes and Design for Processes, *edited by Edward Miller*
9. Turbomachinery: Basic Theory and Applications, *by Earl Logan, Jr.*
10. Vibrations of Shells and Plates, *by Werner Soedel*
11. Flat and Corrugated Diaphragm Design Handbook, *by Mario Di Giovanni*
12. Practical Stress Analysis in Engineering Design, *by Alexander Blake*
13. An Introduction to the Design and Behavior of Bolted Joints, *by John H. Bickford*
14. Optimal Engineering Design: Principles and Applications, *by James N. Siddall*
15. Spring Manufacturing Handbook, *by Harold Carlson*
16. Industrial Noise Control: Fundamentals and Applications, *edited by Lewis H. Bell*
17. Gears and Their Vibration: A Basic Approach to Understanding Gear Noise, *by J. Derek Smith*
18. Chains for Power Transmission and Material Handling: Design and Applications Handbook, *by the American Chain Association*

19. Corrosion and Corrosion Protection Handbook, *edited by Philip A. Schweitzer*
20. Gear Drive Systems: Design and Application, *by Peter Lynwander*
21. Controlling In-Plant Airborne Contaminants: Systems Design and Calculations, *by John D. Constance*
22. CAD/CAM Systems Planning and Implementation, *by Charles S. Knox*
23. Probabilistic Engineering Design: Principles and Applications, *by James N. Siddall*
24. Traction Drives: Selection and Application, *by Frederick W. Heilich III and Eugene E. Shube*
25. Finite Element Methods: An Introduction, *by Ronald L. Huston and Chris E. Passerello*
26. Mechanical Fastening of Plastics: An Engineering Handbook, *by Brayton Lincoln, Kenneth J. Gomes, and James F. Braden*
27. Lubrication in Practice, Second Edition, *edited by W. S. Robertson*
28. Principles of Automated Drafting, *by Daniel L. Ryan*
29. Practical Seal Design, *edited by Leonard J. Martini*
30. Engineering Documentation for CAD/CAM Applications, *by Charles S. Knox*
31. Design Dimensioning with Computer Graphics Applications, *by Jerome C. Lange*
32. Mechanism Analysis: Simplified Graphical and Analytical Techniques, *by Lyndon O. Barton*
33. CAD/CAM Systems: Justification, Implementation, Productivity Measurement, *by Edward J. Preston, George W. Crawford, and Mark E. Coticchia*
34. Steam Plant Calculations Manual, *by V. Ganapathy*
35. Design Assurance for Engineers and Managers, *by John A. Burgess*
36. Heat Transfer Fluids and Systems for Process and Energy Applications, *by Jasbir Singh*
37. Potential Flows: Computer Graphic Solutions, *by Robert H. Kirchhoff*
38. Computer-Aided Graphics and Design, Second Edition, *by Daniel L. Ryan*
39. Electronically Controlled Proportional Valves: Selection and Application, *by Michael J. Tonyan, edited by Tobi Goldoftas*
40. Pressure Gauge Handbook, *by AMETEK, U.S. Gauge Division, edited by Philip W. Harland*

OTHER VOLUMES IN PREPARATION

PRESSURE GAUGE HANDBOOK

U.S. GAUGE DIVISION
SELLERSVILLE, PENNSYLVANIA

edited by Philip W. Harland
Engineering Consultant

Marcel Dekker, Inc.   New York and Basel

Library of Congress Cataloging-in-Publication Data
Main entry under title:

Pressure gauge handbook.

(Mechanical engineering ; 40)
Includes index.
1. Pressure-gages—Handbooks, manuals, etc.
I. Harland, Philip W., [date]. II. Ametek, Inc.
U.S. Gauge Division. III. Series.
TJ370.P74 1985   621.1'97   85-13054
ISBN 0-8247-7433-7

COPYRIGHT © 1985 by AMETEK, INC. ALL RIGHTS RESERVED.

Neither this book nor any part may be reproduced or transmitted in any form or by any means, electronic or mechanical, including photocopying, microfilming, and recording, or by any information storage and retrieval system, without permission in writing from AMETEK, Inc.

MARCEL DEKKER, INC.
270 Madison Avenue, New York, New York 10016

Current printing (last digit):
10 9 8 7 6 5 4 3 2 1

PRINTED IN THE UNITED STATES OF AMERICA

Foreword

The need for a convenient and reliable means to measure pressure originated with the Industrial Revolution, and the need continues today. Early devices consisted of various arrangements of spring-loaded pistons and cylinders, but these were subject to leakage and excessive friction between moving parts.

The discovery that pressure applied to a length of coiled tubing having a flattened cross section will cause the tubing to uncoil in a manner which is proportional to the applied pressure made it possible to produce a leakfree, frictionless, pressure-sensitive element. The element has a high degree of reliability, can be mass produced at a low cost, and may be used for pressures as low as one atmosphere and as high as 10,000 atmospheres. The element motion can be mechanically amplified, without the addition of any external power source, to give an accurate and highly visible indication of the applied pressure. The discovery was so basic that it is still used today.

As noted in the text, the uses and variety of pressure sensors are myriad, and literally millions of them are produced annually. Despite the fact that almost every household contains several pressure sensors (domestic heating systems, domestic well water supplies, automobile panels, fire extinguishers, etc.) and an industrial manufacturing plant may contain hundreds, many of their essential characteristics are unknown, even by those people who specify their use and procurement. This handbook presents a wealth of information for anyone who needs to know more about pressure sensors.

R. L. Noland
President, AMETEK, Inc.

Preface

This handbook presents a detailed discussion of pressure gauges of the type which contains various types of elastic measuring elements (principally bourdons) wherein the power necessary to provide an analog indication of the magnitude of the applied pressure is supplied by the applied pressure. Pressure gauges requiring an external source of energy (e.g., strain gauges, and piezoresistive, capacitive, and variable reluctance transducers) are considered only briefly.

The work is intended as a reference book for anyone who specifies, installs, procures or maintains pressure gauges. With this in mind, the technical information presented has been selected so as not to burden the reader with material which would be of interest only to pressure gauge designers, yet be sufficient to give the background necessary to make informed choices with respect to the specification, installation, procurement and maintenance of pressure gauges, and related accessories. Therefore, the major portion of the text presents the many options available from the manufacturer, and the advantages and limitations of these options from the standpoint of performance and cost. Illustrations are used extensively to provide additional clarity to the text.

Over the 100 years or so that pressure gauges have been mass produced, a sometimes bewildering array of construction variations, nomenclature, and accessories has evolved. Chapters 2, 3 and 4 are intended to clear up at least some of the confusion by discussing and illustrating the more common variations and defining the nomenclature. Chapter 5 continues this objective by discussing a variety of accessories available for use with pressure gauges which can improve their performance and expand their usefulness. Pressure gauge selection, installation, and maintenance are covered in Chapters 6 and 7, including detailed information on how to use the various built-in adjustments for recalibration of a pressure gauge to restore its accuracy. Chapter 8 discusses the several ways that pressure gauges may be used to measure temperature. Chapter 9 gives a general review of externally powered pressure transducers and is included in order to give the reader some direction in the event self-powered gauges, which are the subject of this book, will not provide the desired performance characteristics.

Throughout this handbook references are made to various safety considerations involved in the use of pressure gauges. Chapter 10 is devoted to this topic. Its importance will become apparent to anyone who works with pressurized liquids and gases. A discussion of the more recent ISO units of pressure and how they relate to previously established metric and English units is contained in Chapter 11.

To sum up, the handbook provides a basic knowledge of the operation, selection and maintenance of pressure gauges, which should enable those who are responsible for these functions to make effective choices with respect to performance requirements, application requirements, and initial cost.

Philip W. Harland

Acknowledgments

The editor gratefully acknowledges the following people for their suggestions and reviews of various portions of the text:

Daniel M. Clementi
 Senior Project Engineer,
 AMETEK, Inc./U.S. Gauge Division

T. A. Stuart Duff
 Chief Engineer, Operations,
 AMETEK, Inc./U.S. Gauge Division

Milton Mollick
 Manager, Special Products Engineering,
 AMETEK, Inc./U.S. Gauge Division

Charles J. Reed
 Director of Engineering,
 AMETEK, Inc./Mansfield and Green Division

The cooperation of the American Society of Mechanical Engineers in granting permission to use the material in Chapter 10 is also acknowledged and appreciated.

Contents

Foreword		iii
Preface		v
Acknowledgments		vii
1	**FUNDAMENTALS OF PRESSURE MEASUREMENT**	1
1.1	Reasons for Measuring Pressure	1
1.2	Application of Pressure Gauges	2
1.3	Pressure Principles	3
1.4	Kinds of Pressure	7
1.5	Atmospheric Pressure	11
1.6	Pressure Units	14
1.7	Gravimetric and SI Units	14
2	**FUNDAMENTALS OF PRESSURE GAUGES**	19
2.1	Introduction	19
2.2	Classification of Pressure Gauges	20
2.3	Classification by Function	21
2.4	Classification by Case Type	27
2.5	Classification by General Field of Use	29
2.6	Classification by Specific Application	30
2.7	Classification by Accuracy	32
2.8	Classification by Type of Measuring Element	37

3 GAUGE COMPONENTS — 51

3.1	Basic Components	51
3.2	Socket	52
3.3	Measuring Element	55
3.4	Bourdons	55
3.5	Metallic Diaphragms	66
3.6	Formed Metallic Bellows	68
3.7	Welded Bellows	70
3.8	Nonmetallic Pressure Elements	71
3.9	Choosing the Measuring Element	71
3.10	Tips	71
3.11	Assembly of Socket, Bourdon, and Tip	74
3.12	Gauge Movements	77
3.13	Dials	97
3.14	Pointers	104
3.15	Case and Ring	108
3.16	Windows	122

4 LIQUID-FILLED GAUGES — 125

4.1	Introduction	125
4.2	Internal Case Pressure	125
4.3	Relieving Internal Case Pressure	127
4.4	Choice of Fill Fluid	129
4.5	Case Venting of Liquid-Filled Gauges	130
4.6	Compatibility of Fill Fluid	130
4.7	Case Style	130
4.8	Pressure Elements	131

5 GAUGE ACCESSORIES — 133

5.1	Introduction	133
5.2	Diaphragm Seals	134
5.3	Case Pressure Relief Devices	142
5.4	Pulsation Dampers — Pressure Snubbers	144
5.5	Gauge Cocks	150
5.6	Siphons	150
5.7	Bleeders	152
5.8	Heaters for Gauges and Connecting Lines	155
5.9	Maximum and Minimum Pointers	156
5.10	Electrical Switches	160
5.11	Shut-off Valves and Gauge Protectors	167

Contents

6	**GAUGE SELECTION AND INSTALLATION**	169
	6.1 Introduction	169
	6.2 Nature of Pressurized Medium	169
	6.3 Environmental Conditions	174
	6.4 Method of Connecting Gauge to Pressurized Medium	176
	6.5 Method of Mounting Gauge	179
	6.6 Required Size of Gauge	180
	6.7 Required Accuracy of Gauge	181
7	**GAUGE MAINTENANCE AND CALIBRATION**	183
	7.1 Introduction	183
	7.2 Conditions Affecting Accuracy and Performance	183
	7.3 Maintenance Program	184
	7.4 Maintenance Facility	184
	7.5 Pressure Standards	186
	7.6 Calibration of Pressure Gauges	196
	7.7 Correction of Errors	200
	7.8 Maintenance of Diaphragm Seals	209
8	**PRESSURE-ACTUATED THERMOMETERS**	215
	8.1 Introduction	215
	8.2 Gas-Filled Thermometers	215
	8.3 Liquid-Filled Thermometers	219
	8.4 Vapor-Pressure Thermometers	221
	8.5 Effect of Shipping Temperature	227
	8.6 Effect of Ambient Pressure	227
	8.7 Effect of Bulb Position	227
9	**PRESSURE TRANSDUCERS**	229
	9.1 Introduction	229
	9.2 Rheostats	230
	9.3 Potentiometers	233
	9.4 Linear Variable Differential Transformers	235
	9.5 Strain Gauges	238
	9.6 Capacitance Transducers	240
	9.7 Accuracy of Pressure Transducers	242
10	**SAFETY**	247
	10.1 Introduction	247
	10.2 Excerpt from ANSI B40.1-1980	248
	10.3 Additional Safety Information	254

	10.4 Failure Modes	254
	10.5 Factors to Be Considered	255
11	METRIC CONSIDERATIONS	259
	11.1 Introduction	259
	11.2 Use of Gravimetric Units	262
	11.3 Negative Pressure	262
	11.4 Foreign Standards	263
12	ORDERING AND SPECIFICATION INFORMATION	265
	12.1 Introduction	265
	12.2 Gauges for In-Plant Use	265
	12.3 Gauges for Use by Original Equipment Manufacturers	266
	12.4 Information Supplied to Gauge Manufacturer	266
APPENDIX: TABULAR DATA		269
INDEX		283

PRESSURE GAUGE HANDBOOK

1
Fundamentals of Pressure Measurement

1.1 REASONS FOR MEASURING PRESSURE

There are many reasons why it may be necessary to measure the pressure of a gas such as air or carbon dioxide, or a liquid such as water or oil. In some applications, only a rough indication of the pressure is needed, while in others the pressure may be critical, requiring an accurate measurement in order to avoid endangering personnel and equipment. The following paragraphs list several reasons for measuring pressure.

1.1.1 Providing Operating Information

Many products from the petrochemical industries require careful control of pressure during the manufacturing process. While the actual control will probably be automatic, a pressure gauge will give the operator a constant indication of the pressure so that adjustments can be made to the control loop.

In some applications the normal pressure of the process may not be highly critical, but it is necessary to know when the pressure exceeds some set limit. If the pressure becomes too high due to an abnormal condition in the process, it may cause damage to instrumentation, pumps, or other equipment or even burst the pressure vessel. A continuous indication of the pressure will allow the operator to shut down the process or vent the system before any damage is done.

1.1.2 Providing Test Data

In the shop or laboratory, pressure measurements are often required as part of the evaluation and testing of materials or equipment. Generally such measurements must be made with a high degree of accuracy, requiring the use of a special class of gauges called test gauges.

1.1.3 Measuring Quantity

The quantity of a gas stored in a tank is proportional to the pressure of the gas. Therefore, a pressure gauge when installed on a tank of known volume can be calibrated in terms of quantity, enabling the user to determine how much gas is consumed by a particular process.

The quantity of liquid in a tank of known volume can be calculated by measuring the pressure of the liquid at the bottom of the tank. Gauges are furnished calibrated in terms of gallons of a specified liquid.

1.1.4 Indicating Operational Readiness

Sprinkler systems for fire protection and fire extinguishers are maintained under pressure, and a pressure gauge is installed in the sprinkler system or on the fire extinguisher to indicate the amount of the pressure. Gauges for this purpose are often supplied with a dial displaying an operating zone so that the user can easily determine if the equipment is in fact operational.

1.1.5 Measuring Force

The force generated by a piston and cylinder can be calculated by multiplying the area of the piston by the pressure acting on the area. Thus gauges for use with hydraulic presses may be supplied with dials calibrated in terms of tons of force.

1.2 APPLICATION OF PRESSURE GAUGES

For the reasons listed above, various types of pressure gauges are found on a wide variety of equipment used in industry and the home. A list of typical equipment includes the following:

Pressure regulators for oxygen, carbon dioxide, acetylene, and other compressed gases

Compressors for air and ammonia
Fire extinguishers
Sprinkler systems for fire protection
Hydraulic presses
Domestic hot-water heating systems
Pressure vessels in the processing industries
Fire trucks
Highway vehicles
Off-highway tractors
Pipelines
Test stands, hydraulic and pneumatic
Domestic well pumps
Fertilizer sprayers
Refrigeration equipment
Automotive test equipment, including tire gauges, compression meters, and manifold vacuum gauges
Air-line lubricators and filters
Vacuum systems
Steam boilers and turbines
Aircraft, including fuel pressure, hydraulic pressure, air-speed, etc.
Oil refinery equipment
Pneumatic data transmitters
Medical equipment
Paint sprayers
Swimming pool filters

1.3 PRESSURE PRINCIPLES

1.3.1 Definition of Pressure

Pressure is defined as a force acting over a given area. In Fig. 1.1, for example, a weight of 1 lb (the force) is shown pressing against water contained in a glass cylinder. Assume the cylinder has an area of 1 in.^{2} and there is no friction between the weight and the wall of the cylinder. The water directly below the weight must be at a pressure of one pound per square inch (abbreviated as 1 psi). As we will see later, the pressure at the bottom of the cylinder is somewhat higher than 1 psi because the weight of the water between the one-pound weight and the bottom of the cylinder also creates pressure.

1.3.2 Liquid Columns: Head Pressure

If the cylinder is connected to a vertical glass tube A as shown in Fig. 1.1, the 1-psi pressure will force the water up the tube. As

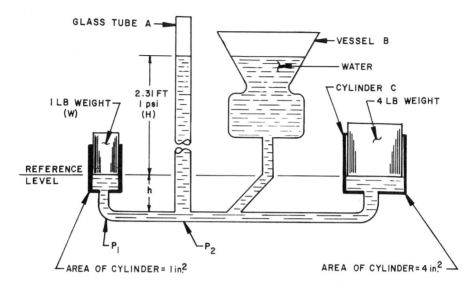

Fig. 1.1 The nature of hydrostatic pressure.

the water rises in the tube, it develops a downward force due to the weight of the water. When this force balances the force of the one-pound weight, the water will stop rising.

The height H to which the water will rise is 2.31 ft above the reference level, which is the level corresponding to the underside of the weight. Note that the length of the piping below the reference level (shown as h) is not important, because the weight of the water from the reference level to point P_1 is exactly balanced by the weight of the water from the reference level to point P_2. The height that a liquid column will rise for a given pressure is proportional to the density (defined as weight per unit volume) of the liquid. If mercury, which is 13.6 times as dense as water, were used in place of water, it would have to rise only 2.31 ÷ 13.6 or 0.17 ft (approximately 2 in.) to balance the 1 psi created by the weight W.

A column of liquid that develops a pressure due to its height is spoken of as a *head* of that liquid. For example, the water in the column of Fig. 1.1 is a 2.31-ft head of water; if the liquid were mercury it would be a 2-in. head of mercury. Observe that the liquid must be specified because of the variations in the density of different liquids.

1.3.3 Calculating Head Pressure

The pressure created by a given head of liquid or, conversely, the head of liquid that will be created by a given pressure can be calculated using the relationship $P = D \times H$, where P is the pressure, D the density of the liquid, and H the head or height of the liquid column.

In making calculations, it is necessary to use "compatible units," that is, if the density is expressed in pounds per cubic inch (lb/in.^{3}), then the head must be in inches (not feet or centimeters), in which case the pressure will be in psi. If the density is expressed in grams per cubic centimeter, then either it must be converted to lb/in.^{3} or the head must be expressed in centimeters, in which case the pressure will be in terms of grams per square centimeter (g/cm^2). Example: Show that 2.31 ft of water is equivalent to 1 psi.

To make the calculation, it is necessary to know the density of water. In the Appendix at the end of this volume, Table A.5 gives the density of water at 20°C as 62.316 lb/ft^3. Since the pressure is to be in terms of psi (lb/in.^{2}), the density must be converted to lb/in.^{3}, which is accomplished by dividing lb/ft^3 by 1728—the number of in.^{3} in 1 ft^3. Also, the 2.31 ft must be converted to inches. Using the relationship given above,

$P = D \times H$

$P = \dfrac{62.316}{1728} \times 2.31 \times 12 = 1.00 \text{ psi}$

Checking the units in the above calculation shows that

$P = \dfrac{lb}{ft^3} \times \dfrac{ft^3}{in.^3} \times ft \times \dfrac{in.}{ft} = \dfrac{lb}{in.^2}, \text{ or psi}$

If the odd-shaped vessel B of Fig. 1.1 is also connected to the glass cylinder, then the height of the water column in vessel B will rise to the same level as in the cylindrical tube A. The odd shape was chosen to illustrate that the area and shape of the column do not affect the height of the head created by the 1-psi pressure. The tube connecting vessel B to the system does not have to be vertical and may take any path, including one with reverse bends, to reach vessel B. Note also that the weight W supports the head of water in both columns A and B. Any number of columns of any size or shape could be added to the system, and the 1-lb weight would support a head of water 2.31 ft high in every column.

1.3.4 Force and Pressure

The relationship between force and pressure is given by the equation

$$F = PA$$

where F is the force, P the pressure, and A the area. The three factors must, of course, be expressed in compatible units.

If cylinder C, shown on the right-hand side of Fig. 1.1, has an area of 4 in.^{2}, then the pressure of 1 psi created by the 1-lb weight W will support a weight operating in cylinder C equal to 4 lb. This is the principle used in hydraulic jacks, wherein a small-diameter piston/cylinder is connected to a large-diameter piston/cylinder. Through some mechanical linkage the operator applies a force to the small piston, creating a pressure equal to the force divided by the area of the piston. This pressure is then applied to the large-diameter piston, creating a force equal to the pressure times the area of the large-diameter piston. For example, if the small piston has an area of 0.05 in.^{2} (approximately 1/4 in. in diameter), then a force of 150 lb applied to it will create a pressure of 3000 psi. If the large piston has an area of 1.3 in.^{2} (approximately 1 1/4 in. in diameter), then a pressure of 3000 psi applied to it will create a force of 3900 lb. From this it will be seen that

$$F_l = F_S \times \frac{A_l}{A_S} = F_S \times \frac{D_l^2}{D_S^2}$$

where F_l is the force on the large piston, F_S the force on the small piston, A_l the area of large piston, A_S the area of small piston, D_l the diameter of large piston, and D_S the diameter of small piston.

In other words, the force applied by the operator is increased by the ratio of the two areas, which is equivalent to saying that the force is increased by the ratio of the square of the two diameters.

It should be apparent that the motion of the large piston will be less than the motion of the small piston by the inverse ratio of the areas; continuing with the above example, the volume displaced by a 1-in. motion of the small piston will move the large piston by only 1/26 in.

It is important to note that if the 1-lb weight of Fig. 1.1 creates a pressure of 1 psi in the apparatus, then the converse is true; that is, a pressure of 1 psi in the apparatus acting over an area of 1 in.^{2} will create sufficient force to support the 1-lb weight. This principle is the basis for the deadweight tester, which is an important means of measuring pressure. For further discussion, see Sec. 7.5.6. It is also true that if a pressure of 1 psi will support a liquid column 2.31 ft high, then the same column will create a pressure of 1 psi in the apparatus. This is the reason for connecting a fire-protection sprinkler system to elevated water tanks; pressure will be maintained in the system at all times, even during a pumping failure.

1.3.5 Manometers for Pressure Measurement

If the 1-lb weight in Fig. 1.1 is replaced by an unknown pressure, then the value of H will be a direct measure of pressure. For example, if the unknown pressure raised the water column so that H equaled 2.3 ft of water, then the unknown pressure must be equal to 1 psi.

Actually, in refined forms, such devices (called manometers) provide an accurate means of measuring pressure. Because water requires a height of 2.31 ft to measure only 1 psi, water is used for relatively low pressures. For higher pressures, mercury is used. Usually, such manometers are graduated in terms of inches of mercury (abbreviated in. Hg). Water manometers are graduated in terms of inches of water (abbreviated in. H_2O). The letters Hg represent the chemical symbol for mercury, and H_2O is the chemical symbol for water. See Secs. 1.7.3 and 7.5.7 for further discussion of liquid columns.

Figure 1.2 shows the relation between several commonly used pressure units. For rough comparison, this shows 1 ft of water and 1 in. of mercury are approximately equal to 1/2 psi.

1.4 KINDS OF PRESSURE

Pressure is measured using various reference points as zero, which results in what may be called different kinds of pressure. The various terms used to describe these are gauge pressure, absolute pressure, negative pressure, differential pressure, and vacuum. Fig. 1.3 illustrates the distinctions between these terms.

1.4.1 Gauge Pressure

The most common type of pressure gauge is constructed so that the measured pressure is applied to the inside of the measuring element (i.e., the bourdon) and atmospheric pressure surrounds the outside of the measuring element. Therefore, the pressure indicated

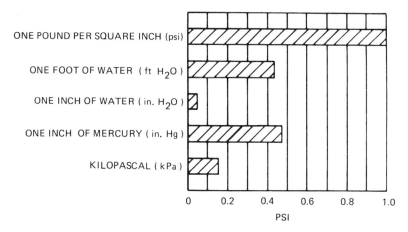

Fig. 1.2 Relationship of commonly used units of pressure.

by the gauge will be the amount that the measured pressure is in excess of the surrounding atmospheric pressure. This kind of pressure is called gauge pressure, and its reference point is the surrounding (ambient) atmospheric pressure. Atmospheric pressure is often referred to as barometric pressure. The left-hand side of the chart in Fig. 1.3 illustrates a gauge pressure of 50 psi. To be strictly correct, this should be written as 50 psig to indicate it is gauge pressure. However, by custom, it may be assumed that the measurement is in terms of gauge pressure unless otherwise stated, and therefore the letter g is usually omitted.

1.4.2 Absolute Pressure

Atmospheric pressure is the result of the weight of the earth's atmosphere and varies depending on the altitude and the prevailing weather. A fixed (unchanging) reference point may be established at an atmospheric pressure equal to zero; this point is called zero absolute pressure. Gauges may be constructed to use zero absolute pressure as a reference point, in which case the indication would be in terms of a unit such as psi absolute, which is abbreviated as psia. Fig. 1.3 introduces the term *standard atmospheric pressure* and equates it to 14.696 psia. This pressure represents the weight of a column of air 1 in.^{2} in cross section having a height that extends from sea level to the outer limits of the earth's atmosphere (see Sec. 1.5.1 for further discussion).

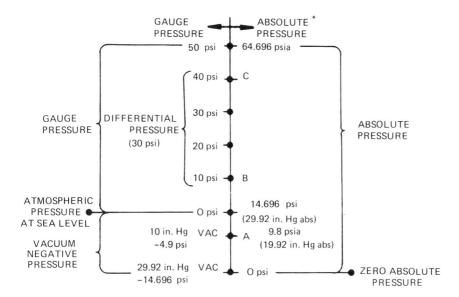

* BASED ON STANDARD ATMOSPHERIC PRESSURE = 14.696 psia
OR 29.92 in. Hg abs

Fig. 1.3 Kinds of pressure.

1.4.3 Negative Pressure

Negative pressure is a term intended to replace *vacuum* (for additional discussion of negative pressure, see Sec. 11.3). Negative pressure, or vacuum, is a pressure below atmospheric pressure and therefore, like gauge pressure, uses the surrounding atmosphere as the reference point. Because of this, the maximum negative pressure or vacuum that can be obtained is a value numerically equal to the existing ambient atmospheric pressure (barometric pressure), at which point the absolute pressure would be zero. It is customary to express vacuum in terms of a liquid head (in Hg, cm H_2O, etc.), whereas negative pressure is usually expressed in force/area units, except that a minus sign is placed in front of the numerical value to indicate that the pressure is negative (-10 psi, -70 kPa, etc.). In both cases the gauge indicates atmospheric pressure as zero, and the numerical value of the indications increases as the negative pressure or vacuum approaches zero absolute pressure. There is, of course, no such thing as negative absolute pressure, since the reference point for absolute pressure is zero atmospheric pressure. A vacuum in a chamber would be indicated if the absolute pressure was less than the barometric pressure. Point A in Fig. 1.3 illustrates that a vacuum of 10 in. Hg and a negative pressure of -4.9 psi are equivalent, and both would be equivalent to an absolute pressure of 9.8 psia if the barometric reading was 14.696 psia (29.92 in. Hg absolute).

1.4.4 Differential Pressure

One other term is illustrated in Fig. 1.3, namely, *differential pressure*. Gauges are available to measure and indicate the difference between two pressures, such as B (10 psi) and C (40 psi) in Fig. 1.3. The differential pressure gauge indicates the difference on a scale that would read zero when the two pressures are equal—regardless of their individual values. The letter d is used to indicate difference, so that in this example the differential pressure between points B and C would be expressed as 30 psid (C minus B). The zero point may be in the center of the scale to permit readings in one direction or the other, depending on which pressure is greater. If one of the two pressures is always the greater of the two, the zero point may be at the beginning of the scale. Because such gauges read the difference between two pressures, the distinction between gauge and absolute pressure is irrelevant; variations in atmospheric pressure will not alter the readings because both inputs are identically affected by barometric changes.

1.4.5 Converting from Gauge Pressure to Absolute Pressure

Gauge pressure can be converted to absolute pressure by adding the gauge pressure reading and the atmospheric pressure at the time the reading was taken. Of course, the units of pressure that are added must be identical. For example, if it is desired to convert a gauge pressure reading of 60 psi, made when the barometric pressure was 30.2 in. Hg, to absolute pressure, it is first necessary to convert the 30.2 in. Hg to psi. This is done by multiplying 30.2 by the factor of 0.491 psi per in. Hg (see Sec. 1.7.4) and adding the result, 14.8 psi, to 60 psi. Thus

$$60 \text{ psi} + \left(\frac{30.2 \text{ in. Hg} \times 0.491 \text{ psi}}{\text{in. Hg}} \right) = 74.8 \text{ psia}$$

If the gauge pressure reading in the above example is -10 psi (negative pressure), the same procedure is followed, taking into account the negative sign. Thus

Atmospheric Pressure 11

-10 psi $+ 14.8$ psi $= 4.8$ psia

If the gauge pressure reading is in terms of vacuum, then it is necessary to subtract the vacuum reading from the atmospheric pressure. Continuing with the same example, if the gauge pressure reading is 12 in. Hg vacuum, simply subtract the reading from the atmospheric pressure of 30.2 in. Hg. Thus

30.2 in Hg - 12 in. Hg vac = 8.2 in. Hg absolute

Note that in this case, the letter a is not used to indicate "absolute" and it is necessary to spell out the word or abbreviate it as abs.

1.5 ATMOSPHERIC PRESSURE

As discussed in Sec. 1.4, the pressure indicated by a gauge that measures in terms of absolute pressure is not affected by changes in atmospheric pressure; such gauges measure the pressure with respect to an unchanging absolute zero reference. Gauges that measure in terms of gauge pressure (or vacuum) are affected by the change in atmospheric pressure, because the pressure element of these gauges senses the difference in pressure between that applied to the interior of the element and that applied to the exterior, which is usually atmospheric pressure. Therefore, a decrease in atmospheric pressure surrounding the element has the same effect as an increase in pressure applied to the interior of the element.

The importance of changes in atmospheric pressure depends on the required accuracy of indication and the magnitude of the pressure being measured. Before considering these points, it is necessary to explain the nature of atmospheric pressure and how it is measured.

1.5.1 Standard Atmospheric Pressure

In Sec. 1.4.2 atmospheric pressure at sea level was referred to as 14.696 psia (equivalent to 760 mm Hg absolute or 29.92 in. Hg absolute). This is actually an average value above zero absolute pressure measured at sea level. It is based on many readings taken under general conditions at various locations. Changes in atmospheric conditions will vary the atmospheric pressure. Measurement of this pressure and how it is changing is one way of predicting changes in the weather. Also, atmospheric pressure decreases as the altitude

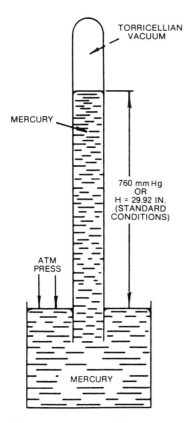

Fig. 1.4 Torricelli's vacuum.

increases. Table A.6, in the Appendix, gives the average atmospheric pressures at various altitudes; note that at 3,000 ft it is 13.178 psia and at 10,000 ft it is 10.108 psia.

1.5.2 Barometers

Measurement of atmospheric pressure is made with a simple instrument known as a barometer (literally, a pressure meter). Fig. 1.4 shows the first barometer, devised in 1643 by Evangelista Torricelli, an Italian physicist. He filled a long glass tube, closed at one end,

Atmospheric Pressure

with mercury. Then he inverted the tube, putting the open end into a dish of mercury; the mercury level dropped in the tube, leaving a space in the top of the tube. This space was known as "Torricellian vacuum."

Science students at that time had been taught that "nature abhors a vacuum"; hence, there should be no tendency for the mercury to drop and leave a vacuum above it. Torricelli's vacuum was, therefore, of great interest. It took nearly five years for this experiment to prove that the column of mercury in the tube was supported by the pressure of the earth's atmosphere acting on the surface of the mercury in the dish. The height of the mercury column at standard atmospheric pressure is 29.92 in. regardless of how much the length of the tube exceeds 29.92 in.

Because of its measurement by such mercury barometers, atmospheric pressure is often referred to as barometric pressure. There are barometers today that measure atmospheric pressure without the use of mercury; one type is called an aneroid (meaning "without liquid"). This instrument uses a diaphragm capsule with all the air removed ("evacuated") as is described in Chap. 2. It is more convenient to use than the mercury barometer. However, mercury barometers are still in common use today because of the accuracy and stability they provide.

1.5.3 Effects of Atmospheric Pressure Changes

Consider now exactly how any change in atmospheric pressure affects the measurement of pressure. Assume we have an absolute pressure gauge (measuring in units of psia) and a standard gauge (measuring in units of psi) attached to a tank. We then charge the tank with air until the standard gauge indicates 6 psi. Since atmospheric pressure surrounds the bourdon of the standard gauge, the pressure in the tank must be 6 psi higher than the atmospheric pressure. The indication on the absolute pressure gauge will be the sum of the atmospheric pressure and the pressure above atmospheric. If the atmospheric pressure happens to be 14.7 psia, then the absolute pressure gauge will indicate 20.7 psia.

Suppose now the atmospheric pressure changes to 14.1 psia. The indication on the standard gauge will increase to 6.6 psi because the pressure surrounding the bourdon decreased by 0.6 psi while the pressure inside the bourdon remained the same. The indication on the absolute pressure gauge will be the sum of the atmospheric pressure and the pressure above atmospheric, or 14.1 + 6.6, which equals 20.7 psia. It will be seen, therefore, that the pressure indicated by the absolute pressure gauge is unaffected by atmospheric pressure changes, whereas the indication of the gauge measuring psi is affected, the amount of the effect being equal to the change in atmospheric pressure.

Because the magnitude of changes in atmospheric pressure at a given altitude is small, the effect on the accuracy of the measurement is significant only when low pressure is being measured. In the example given above, the change in indication was 0.6 psi out of 6.0 psi, or 10%. If the tank had been charged to a pressure of 600 psi, then the change in indication due to the atmospheric pressure change would be 0.6 psi out of 600 psi, or only 0.1%, which is usually not significant. The larger effect of changes in altitude will have greater significance.

1.6 PRESSURE UNITS

As discussed in Sec. 1.3.1, pressure is defined as a force per unit area. Therefore any combination of a unit of force divided by a unit of area will result in a unit of pressure. Thus we have pounds per square inch, pounds per square foot, ounces per square inch, and kilograms per square centimeter. Another group of units is based on the height of a liquid column that can be supported by the pressure. These units are inches of water, inches of mercury, millimeters of mercury, etc.

1.7 GRAVIMETRIC AND SI UNITS

1.7.1 The SI System

All the units discussed in Sec. 1.6 are based on the earth's gravity acting on a mass to create the force. Unfortunately, the magnitude of the earth's gravity varies throughout the world by approximately 0.5%, so a given mass will create a different force depending on its location. If a pressure gauge is calibrated against a liquid column or a deadweight tester at one location, it will indicate a different pressure when checked against the same standard at a location where the gravity is different. Therefore it is necessary to select some value of gravity as standard. This value is 9.80665 meters per second per second (32.174 feet per second per second), and, precision deadweight testers and mercury columns used in locations having a different gravity are "corrected for local gravity."

Units of pressure such as those described in Sec. 1.6 are often called gravimetric units because their definition includes a value for the magnitude of the earth's gravity, usually the standard gravity of 9.80665 meters per second per second.

Gravimetric and SI Units

In 1960, at a meeting of the General Conference on Weights and Measures (CGPM), a system of metric units was formally given recognition by 36 countries, including the United States. This system, called Le Système International d'Unités (abbreviated as SI), designates and defines the unit of linear measure as the meter, the unit of mass as the kilogram, and the unit of time as the second. It is important to note that the kilogram is defined as the unit of mass. Mass is the property of matter that defines its inertia, that is, its resistance to a change in velocity or direction of travel. If a mass is at rest it will have a certain resistance to being put into motion (accelerated). If it is moving at a constant velocity it will move in a straight line and have a certain resistance to a change in its velocity (acceleration or deceleration) or a change in its direction of motion. The mass of a given lump of matter is therefore a constant and is entirely independent of gravity. It has the same mass at any location on the earth or in space. A platinum–irridium cylinder kept at the International Bureau of Weights and Measures in Paris is designated as having a mass of 1 kg.

1.7.2 The Pascal

In the SI system, the unit of force is a derived unit called the newton and is defined in terms of the three basic SI units as that force which will accelerate a mass of 1 kg at the rate of 1 meter per second per second. The unit of pressure is then defined as 1 newton/m^2, which has been designated as a pascal (Pa). It is apparent, therefore, that the pascal is not a gravimetric unit, since the force of gravity does not enter into its definition.

The pascal is actually a very small pressure (0.000145 psi), and it is customary to apply the metric prefix kilo (k), meaning times 1000, which then results in the kilopascal, kPa. For higher pressures the metric prefix mega (M), meaning times 1 million, is sometimes used, which results in the MPa. However, for industrial pressure measurement applications, the use of MPa is discouraged because of the possibility that those not completely familiar with the SI system may not readily distinguish between two gauges, each of which has a dial reading zero to 100, for example, but one being in kPa units and the other in MPa units.

One other unit of pressure often used in the SI system is called the bar. This is an exact conversion of kPa, and therefore, like the kPa, it is not a gravimetric unit. One bar is equal to 100 kPa and has a value nearly that of 1 atm. For additional discussion of SI pressure units, see Chap. 11.

Conversion tables giving the relationship between kPa and gravimetric units are shown in Table A in the Appendix. However, it must be remembered that these conversions are based on the standard gravity of 9.80665 meters per second per second.

1.7.3 Liquid Head Units of Pressure

While pressure units utilizing liquid column height as a measure of pressure are gravimetric units and therefore not recognized in the SI system, it will probably be some time before their use is discontinued. Therefore it will be well to discuss them further. When pressure is expressed in terms of liquid head, either English or metric units of length may be used. Also, because different liquids have different densities (weight per unit volume), the liquid must be named when expressing pressure as a liquid head. Inches, feet, centimeters, and millimeters are the usual units of length. In tables or specifications, combinations of these units for liquid head are frequently abbreviated as follows:

Liquid head	Abbreviation
Inches of water	in. H_2O
Feet of water	ft H_2O
Inches of mercury	in. Hg
Centimeters of water	cm H_2O
Millimeters of mercury	mm Hg

The density of a liquid varies somewhat with temperature. Hence, to completely define pressure in terms of liquid head, the temperature of the liquid must be standardized. The standard temperature for water is usually 20°C (68°F). For mercury, the standard temperature is 0°C (32°F). However, readings of liquid heads at other than these standard temperatures are seldom significantly in error, although more accurate readings may be necessary for the most exacting test or scientific work.

1.7.4 Conversion Between Different Pressure Units

Table A.1 in the Appendix gives the conversion factors for changing a pressure value in one unit to its equivalent in another unit. For example, to convert 29.6 in. Hg to psi, follow these steps:

1. Move down to the left-hand column of the chart and locate "Inches mercury."
2. Move across the chart to the column headed "psi" (column 1).

Gravimetric and SI Units

3. Read 0.49115.
4. Multiply 29.6 in. Hg by 0.49115 and get 14.53804 psi. This value should be rounded to 14.5 psi, since the converted value should not have more significant digits than the value from which it was converted.

To convert 151 in. H_2O to bar, follow these steps:

1. Move down the left-hand column of the chart and locate "Inches water."
2. Move across chart to the column headed "bar" (column 3).
3. Read 0.0024864.
4. Multiply 151 in. H_2O by 0.0024864 and get 0.3754464 bar. This value should be rounded to 0.375 bar, since the original value contains only three significant digits.

To convert 5000 psi to atmospheres, follow these steps:

1. Move down the left-hand column of the chart and locate "Pounds per sq. in."
2. Move across the chart to the column headed "atm" (column 4).
3. Read 0.068046.
4. Multiply 5000 psi by 0.068046 and get 340.23 atm. Rounding this value requires knowing the precision of the 5000 psi. If all the zeros are significant, that is, the reading was not 4999 or 5001, then the conversion should be rounded to 340.2 atm. If only two zeros are significant, that is, the reading was rounded to the nearest 10 psi (larger than 4995 and smaller than 5005), then the conversion should be 340 atm.

2
Fundamentals of Pressure Gauges

2.1 INTRODUCTION

For well over a century, gauge manufacturers have been developing different types of pressure gauges to meet specific user needs. This evolution, combined with the use of different trade names and even different nomenclature for the same type of gauge, has created a bewildering array of terms. Yet it is obviously important that the proper terminology be used in order to avoid, as far as possible, confusion and misunderstandings in both written and spoken communications about gauges.

In this and following chapters we will attempt to introduce the reader to the terminology by discussing the various types of gauges in common use and identifying the various components of a typical pressure gauge and related accessories. Other chapters will cover selection, calibration, and maintenance, where additional terminology will be presented.

All the pressure gauges covered in this handbook indicate the pressure with a pointer moving over a graduated dial. The case or enclosure is usually, but not necessarily, round. The pointer usually turns about a point at the center of the indicating dial, called concentric type, but may be off center, called eccentric type (see Fig. 2.1). A suitable fitting projects from the case for connecting the gauge to the pressure source.

One other characteristic is common to all gauges covered in this handbook; the pressure to be measured is the only source of power

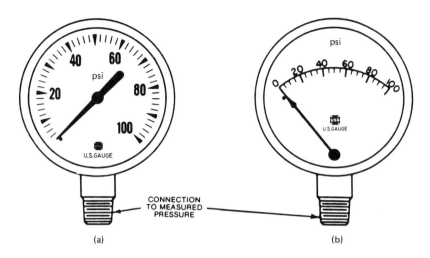

Fig. 2.1 Concentric (a) and eccentric (b) scales.

required to provide the indication. Some form of elastic chamber inside the gauge case converts the pressure to motion, which is translated through suitable links, levers, and gears into movement of the pointer over the indicating scale.

The American Society of Mechanical Engineers sponsored a committee that formulated standards for pressure and vacuum gauges. These standards have been accepted by the American National Standards Institute (ANSI) and are entitled *Gauges — Pressure Indicating Dial Type — Elastic Element*, ANSI B40.1 and ANSI B40.1M-1979. These standards establish a series of standard case sizes, mounting dimensions, ranges, and dial graphics, and provide a section wherein standard gauge terminology is defined. It is strongly recommended that anyone using or specifying pressure gauges obtain a copy of these standards from the American Society of Mechanical Engineers, United Engineering Center, 345 East 47th St., New York, NY 10017.

2.2 CLASSIFICATION OF PRESSURE GAUGES

There are many characteristics by which gauges can be broadly classified. For example:

Function
Case type

Classification by Function 21

General field of use
Specific application
Accuracy
Type of measuring element

The following paragraphs will describe the characteristics of each of the above classifications.

2.3 CLASSIFICATION BY FUNCTION

Pressure gauges are manufactured to perform a variety of functions. The most common is the indication of positive pressure from a single source, using ambient pressure as the zero base; that is, the indication is in terms of gauge pressure (see Sec. 1.4.1).

A gauge that performs this function is most often referred to as simply a pressure gauge. If the gauge performs another function, for example the indication of vacuum or absolute pressure, then it is most often referred to by the function as well. Continuing with the above example, we can have a vacuum gauge or an absolute pressure gauge. In summary, the term *pressure gauge*, used without further qualification, generally refers to a gauge that indicates positive gauge pressure.

It should be noted, however, that *pressure gauge* may be used in the broadest sense, in which case it refers to all types. For example, in the statement "The accuracy of a pressure gauge is affected by ambient temperature changes," it is obvious that all types are included, not just those indicating positive gauge pressure.

2.3.1 Pressure Gauge

A pressure gauge indicates the pressure of a single source in terms of gauge pressure, that is, uses ambient pressure as datum (zero) (see Fig. 2.1).

2.3.2 Vacuum (Negative Pressure) Gauge

A vacuum gauge indicates negative pressure, usually having a span of 0 to 30 in. Hg, which is approximately 1 atm and therefore the maximum vacuum that can be attained (see Fig. 2.2). Traditionally, the bourdon and mechanism are arranged so that the pointer moves clockwise for increasing vacuum in the same manner as a pressure gauge. However, recent standards issued by the International Standards Organization (ISO) and the latest issue of B40.1 (see Sec. 2.1) recommend that the zero vacuum graduation be on the right side

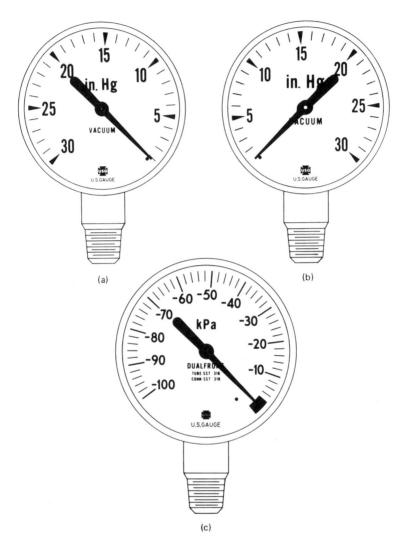

Fig. 2.2 Vacuum gauges. (a) B 40.1 recommended presentation, (b) Traditional presentation, (c) negative pressure.

of the dial and that the pointer move counterclockwise for increasing vacuum and increasing negative pressure. Therefore a gradual changeover to this type of dial presentation can be expected (see Sec. 11.3 for additional discussion of negative pressure).

Classification by Function

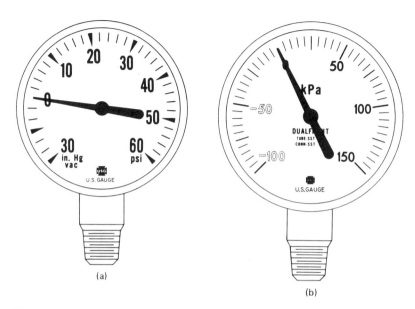

Fig. 2.3 Compound gauges. (a) With vacuum scale, (b) with negative pressure scale.

2.3.3 Compound Gauge

A compound gauge is capable of indicating pressure above and below ambient pressure, that is, positive and negative pressure. The positive pressure portion of the scale is usually calibrated in psi and the negative pressure or vacuum portion in psi or inches of mercury (in. Hg). Thus we might have a 30−0−60 compound gauge, which means the gauge is capable of measuring negative pressure (vacuum) to 30 in. Hg and positive pressure to 60 psi. The negative pressure span is almost always equivalent to 30 in. Hg (100 kPa) regardless of the positive pressure span (see Fig. 2.3).

2.3.4 Duplex Gauge

A duplex, or dual, gauge employs two independent elastic chambers, each connected to a different pressure source. There are two (usually concentric) indicating pointers operating over one dial – each pointer being actuated by its own elastic chamber, entirely independent of the other (see Fig. 2.4). Do not confuse this gauge with the dual scale type defined in Sec. 2.3.6. In use, be sure to check which pointer is operated by which connection.

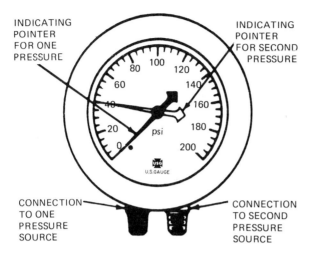

Fig. 2.4 Duplex, or dual, gauge.

2.3.5 Differential-Pressure Gauge

A differential pressure gauge employs two pressure chambers, each connected to a different pressure source. It has one indicating pointer operating over the dial to show the difference in pressure between the two sources.

2.3.6 Dual Scale

A dual scale gauge has a single pressure element, the dial of which contains a basic pressure scale and one or more additional concentric scales graduated in equivalent values of a different pressure unit or other parameters related to the basic pressure, e.g., gas quantity, force, temperature (see Fig. 2.5).

2.3.7 Retard Gauge

In a retard gauge, the measuring element moves freely through only a portion of its pressure span, usually the lower portion. At a definite point in the upper portion of its pressure span, the rate of the remaining motion per unit of pressure change is reduced. Fig. 2.6 shows a typical retarded scale; note that the graduations are uniform to 80 psi, which is almost at 90% of full scale. From

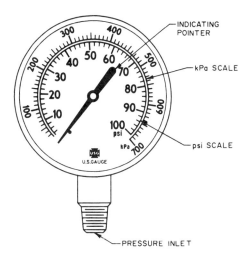

Fig. 2.5 Dual scale gauge.

80 psi to 250 psi, the dial is not graduated and the total scale length in this portion equals approximately that of 10 psi on the lower portions. Such a gauge is referred to as "0 to 250 psi; retarded from 80 psi to 250 psi." However, to completely describe a retard gauge, the portion of the scale length that is nonretarded must be stated. In the example given, this is approximately 240°.

2.3.8 Suppressed Scale Gauge

A gauge having a suppressed scale shows only the upper portion of the total pressure range. At pressures below the minimum shown on the scale, the indicating pointer is not actuated by the elastic chamber and remains inactive at the minimum point of scale until the pressure rises above the minimum value.

A special gauge called a receiver gauge is an example of one having a suppressed scale. In many petrochemical manufacturing processes, operational parameters such as pressure, temperature, and flow are transmitted from the point of measurement to a central control station, using pneumatic pressure as the means of transmission. A transmitter installed at the point of measurement measures the parameter of interest and converts it to a proportional pneumatic pressure, usually between 3 and 15 psi. For example, if it is

Fig. 2.6 Retard gauge.

desired to measure and transmit a flow between 0 and 100 gallons per minute, the transmitter is designed so that at 0 flow its output signal will be 3 psi and at 100 gpm its output will be 15 psi. These pressures are piped to the receiver gauge located on the control panel. The receiver gauge is designed with a slip link (see Sec. 3.12.6 and Fig. 3.26), so that there will be no pointer motion until a pressure of 3 psi is reached. The pointer will continue to traverse the scale, and at 15 psi it will be at its maximum travel, usually 270 degrees from the 3-psi position. The dial of the receiver gauge may be calibrated in any parameter using the 3- and 15-psi reference points. The dial of the gauge shown in Fig. 2.7 illustrates a typical receiver gauge dial. Note that the 0 graduation of the scale corresponds to the 3-psi pip and the 1.0 graduation corresponds to the 15-psi pip. The interval between each of the pips around the dial represents 1 psi. In this instance the actual parameter being measured is masked by providing a scale of 0 to 1.0. This is sometimes done to permit the gauge to be used to indicate a variety of parameters without the necessity of changing dials. Operating instructions would be simply to maintain the process between 0.5 and 0.6, for example.

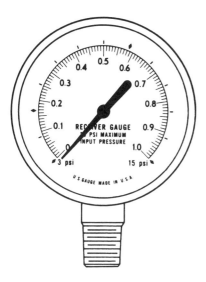

Fig. 2.7 Suppressed scale gauge.

2.4 CLASSIFICATION BY CASE TYPE

Pressure gauges are supplied in a wide variety of case types and are described by such characteristics as size, method of mounting, location of pressure connection, and general construction of case and ring.

2.4.1 Size

In the United States the preferred case sizes are 1 1/2, 2, 2 1/2, 3 1/2, 4 1/2, 6, 8 1/2, 12, and 16 in. in diameter. These diameters refer to the approximate inside diameter of the case at the dial level. There are, of course, variations for specific applications. For example, gauges having square cases, fan-shaped cases, and edge-reading dials are available. Chapter 3 will describe case styles in more detail. International standards, based on metric units, specify standard sizes as 40, 50, 63, 100, 160, and 250 mm. These dimensions refer to the outside diameter of the case assembly.

2.4.2 Method of Mounting

Several common methods of mounting are used. These include:

1. Stem mounting — The gauge is supported by screwing the pressure connection into a mating fitting on the pressurized vessel, leaving the gauge free-standing.
2. Surface mounting — An integral flange at the rear of the case is fastened to a wall or panel. In this instance the entire gauge protrudes in front of the mounting surface.
3. Flush mounting — An integral flange at the front of the case is fastened to a panel. In this instance the major part of the gauge protrudes behind the panel. Often such gauges have a removable ring and glass assembly to permit easy access to a pointer-adjustment feature. The pressure connection is usually on the back of the case.
4. Clamp mounting — A U-clamp installed behind the panel pulls the gauge into the panel. This arrangement is frequently used in automotive panels. The pressure connection is on the back of the case (see Fig. 3.47).

2.4.3 Location of Connection

The pressure connection, in most instances, projects radially from the bottom of the case at the 6 o'clock position, or from the back of the case on the geometrical center, or below the center (see Fig. 3.2). On special order, the gauge can be made with the pressure connection protruding radially from any position around the case, for example, from the 3, 9, or 12 o'clock position. Connections are generally provided with a pipe thread, either 1/8, 1/4, or 1/2 NPT. Special threads can be ordered.

2.4.4 Case and Ring Construction

The two broad categories in this classification are cast (or molded) cases and drawn, sheet metal cases. Generally the lowest cost gauges are made using drawn, sheet metal cases with friction, slip, or threaded rings. (The various ring types will be discussed in more detail in Chap. 3.) Higher-accuracy gauges, such as grade 2A and up, usually have more sturdy cases of molded plastic or cast metal, with threaded rings, bayonet rings or hinged rings. Cast metal cases are most commonly aluminum, and plastic cases are usually molded from a high-impact compound.

The use of plastic cases is growing more common, and they are being used for gauges of all grades of accuracy. Acetal, phenolic,

Classification by General Field of Use 29

polypropylene, and polyester materials are used. The cases may also be "solid front" or "solid back" design. These terms are more fully explained in Chap. 3.

2.5 CLASSIFICATION BY GENERAL FIELD OF USE

Broad terms such as *commercial*, *industrial*, *process*, and *test* have come to be commonly used to classify gauges. Such names are sometimes misleading. It is therefore important to understand the broad intent of these terms.

2.5.1 Commercial Gauges

Gauges found on much of the equipment used in manufacturing plants, stores, garages, etc. are classified as commercial gauges or general purpose gauges. Typical equipment uses include refrigeration units, pumps, compressors, and fire extinguishers. In such applications, although the gauges may be ruggedly constructed, service conditions are not expected to be severe. Such gauges are usually the "drawn case" type, with grade B accuracy (see Sec. 2.7 for an explanation of the various grades of accuracy).

Commercial gauges are generally designed for low unit cost without refinements to simplify maintenance or repair, on the assumption that it is less expensive to replace them than to attempt repairs.

2.5.2 Industrial Gauges

Gauges used in industrial plants for such purposes as measuring steam, oil, and water line pressures are classified as industrial gauges. In general, such gauges have cast cases for additional ruggedness, meet grade A standards of accuracy, and have limited adjustments for repair or maintenance.

2.5.3 Process Gauges

Gauges used in industrial plants for measuring pressures in process equipment, such as in autoclaves, pressure vessels, and process pipelines, are classified as process gauges. Oil refineries, chemical plants, and power plants are typical industries that use many such gauges. The gauges usually have a molded or cast case, meet grade 2A accuracy, and provide for repair or maintenance. They are designed for longer life and reliability and are more expensive than the commercial or industrial grades.

There is no sharp line separating such broad classification terms as *industrial*, *process*, etc. Gauges with grade A accuracy and certain selected design features may well meet the needs of a process application. Similarly, the so-called process gauge with grade 2A accuracy can often be found on what is known as test equipment, such as fuel flow test stands or hydraulic test stands, where an accuracy of ±1/2% is adequate.

2.5.4 Test Gauges

The term *test gauge* is generally applied to gauges having an accuracy of grade 3A or higher. They usually have other features, such as lighter moving parts and smaller bearings in the movement in order to reduce friction and increase sensitivity to small pressure changes. They may be equipped with dials having a mirror band around the periphery to reduce parallax. Grade 4A gauges usually utilize movements with temperature-compensating linkages to minimize calibration shifts resulting from dimensional changes in the various components and the spring rate change of the bourdon.

2.6 CLASSIFICATION BY SPECIFIC APPLICATION

2.6.1 Ammonia Gauges

A number of gauge names have come into use because they refer to a specific application that requires certain design features. One example is the ammonia gauge, which has steel or stainless steel internal parts to withstand the corrosive effects of ammonia. It also has a dial with two scales, one for the pressure and the other for the corresponding temperature of saturated ammonia. The ANSI standards recognize this specific type, specifying that the gauge shall have "plainly inscribed" on its dial the word "ammonia."

2.6.2 Refrigeration Gauges

Refrigeration gauges are designed for service in refrigeration and air conditioning applications and are intended for use with freon. The dual scale usually has pressure graduations in black and the corresponding saturated refrigerant temperature in red. Based

Classification by Specific Application

on servicemen's needs, such gauges often have a convenient zero adjustment and a removable window to provide access to the adjustment.

2.6.3 Oxygen Gauges

These gauges are specifically designed to measure oxygen pressure and must be free of contaminants, which may react with the oxygen and cause an explosion. Various degrees of cleanliness may be specified, but generally level IV, as described in the previously referenced ANSI-B40.1 standard, is used. This requirement limits the number and size of fibers and particles and the maximum quantity of hydrocarbons that may exist within the pressure-containing envelope of the gauge. Anyone involved in the use of gauges to measure oxygen pressure should read the subject matter presented in ANSI-B40.1 for additional information on cleanliness so as to be aware of this important subject.

Gauges used for oxygen service have the dial imprinted with "Use No Oil" in red letters. This indicates the gauge was clean to at least level IV when it was manufactured, and cautions the user not to contaminate the gauge by using it to measure oil pressure, or by using oil or grease on the fitting. Users are cautioned to be aware of the possibility that a gauge whose history of use is unknown will be unsuitable for oxygen service even if the legend "Use No Oil" appears on the dial.

Gauges for use with high-pressure oxygen (1500 psi and over) are generally fitted with a flow restrictor in the inlet port (see Sec. 5.4.4). The purpose of this restrictor is to avoid rapid compression of the gas within the gauge when the gauge is pressurized. Compressing a gas causes the temperature to rise. If the compression is rapid, the temperature increase can be great enough to ignite a contaminant which, in the presence of the oxygen, can cause violent burning or an actual explosion of the gauge. *Removing or altering the restrictor will create a potential hazard to personnel and property.* See Chapter 10 for additional information on the use of pressure gauges with oxygen.

2.6.4 Hydraulic Gauges

Hydraulic gauges are specifically constructed for high-pressure service where water or a noncorrosive liquid is the pressure medium, such as for hydraulic presses. A special link, designed to protect the gauge mechanism against a sudden release of pressure, is often used.

2.6.5 Hydrostatic Head Gauges

Hydrostatic head gauges employ one or more elastic chambers and differ from ordinary types only in the graduation of the scale. The scale is usually graduated to show the head of water, or other liquid, in feet. It may also have a second scale reading in psi. The gauge is used in hot-water heating systems, for example, to show the head of water in the system.

2.6.6 Other Specific Uses

Other gauge names that are self-descriptive of their service include boiler, welding, and sprinkler. Generally, the gauges have been specially designed, or modified from a standard type, to meet specific codes or specifications.

2.7 CLASSIFICATION BY ACCURACY

2.7.1 Meaning of Accuracy

The accuracy of a pressure gauge may be expressed as "percent of span" or "percent of indicated reading." Percent of span is by far the most common method. Use of percent of indicated reading is usually limited to precision test gauges, and unless specifically called out, it may be assumed that an accuracy of ±1/2% means ±1/2% of span. In order to explain each of these methods we must first define span and accuracy.

Span is the algebraic difference between the limits of the unretarded portion of the pressure gauge scale. For example, a pressure gauge manufactured to indicate from 0 to 200 psi over a linear, unretarded scale will have a span of 200 psi.

If the gauge is a retard gauge (Sec. 2.3.7), the span is equivalent to the unretarded portion of the scale. For example, the retard gauge shown in Fig. 2.6 has a span of 80 psi or 550 kPa.

If the gauge is a suppressed scale gauge (Sec. 2.3.8), the span is the difference between the maximum and minimum scale pressure. For example, the suppressed scale gauge shown in Fig. 2.7 has a span of 12 psi.

The span of a compound gauge (Sec. 2.3.3) is the algebraic difference between the positive and negative pressure portions of the scale. In determining the span of a compound gauge, the pressure units of both scales must be the same. For example, the span of the gauge shown in Fig. 2.3a is 75 psi.

Accuracy is defined as the quality of exactness or correctness of the pressure indication. It is measured and graded by the magnitude of the error between the indicated pressure and the true pressure acting on the gauge.

If the accuracy is given as percent of span, then the maximum error that is permitted is determined by multiplying the span by the accuracy percentage and dividing by 100. For example, if a gauge having a span of 300 psi is said to have an accuracy of 1.0%, then the maximum error permitted is

$$300 \text{ psi} \times \frac{1}{100} = 3 \text{ psi}$$

Alternatively, a deviation between the true pressure and the indicated pressure of 1.5 psi anywhere within the 0–300 psi range would be termed a 0.5% error.

Unless otherwise specified, when the accuracy is given as a percent of span, it means plus or minus that percentage. In other words, the indicated pressure may be higher or lower than the true pressure.

If accuracy is given as a percent of indicated reading, then the maximum error that is permitted is determined by multiplying the reading by the accuracy percentage and dividing by 100. For instance, if a gauge having a span of 300 psi is said to have an accuracy of 0.5% of indicated reading, then the maximum error permitted at 300 psi would be

$$300 \text{ psi} \times \frac{0.5}{100} = 1.5 \text{ psi}$$

At an indication of 200 psi, the maximum error permitted would be

$$200 \text{ psi} \times \frac{0.5}{100} = 1.0 \text{ psi}$$

and at 80 psi it would be

$$80 \text{ psi} \times \frac{0.5}{100} = 0.4 \text{ psi}$$

It will be noted that as the indicated pressure decreases the permissible error gets smaller, so it is necessary to set some low limit. Usually this is chosen so that for pressures below 10% of full scale the permissible error remains constant and equal to the permissible error at 10% of full scale. For additional discussion of the terms *accuracy* and *error*, see Secs. 6.7 and 7.6.

	Permissible error ±% of span		
Grade	First 25% of span	Middle 50% of span	Last 25% of span
4A	0.1	0.1	0.1
3A	0.25	0.25	0.25
2A	0.5	0.5	0.5
A	2.0	1.0	2.0
B	3.0	2.0	3.0
C	4.0	3.0	4.0
D	5.0	5.0	5.0

Fig. 2.8 Grades of accuracy as defined in ANSI B 40.1.

The American National Standards B 40.1 and B 40.1M referred to in Sec. 2.1 classify the types of gauges covered in this manual into seven grades of accuracy (see Figs. 2.8 and 2.9). The following is a description of each of these grades and their intended use.

2.7.2 Grade 4A Gauges

Grade 4A gauges offer the highest accuracy covered by the standards, that is, ±0.1% of span over the entire range of the gauge. These gauges are generally made in 12- or 16-in. dial sizes. Smaller sizes do not have sufficient scale length to enable the user to consistently read to an accuracy of 0.1% (one part in 1000). A 12-in. gauge will have a scale length of approximately 26 in. Therefore, the motion of the pointer at its tip for 0.1% of the pressure range would be only 0.026 in. These high-accuracy gauges are usually temperature-compensated. They must be handled carefully in order to retain the accuracy. When using low-range gauges with liquid as the pressure medium, the effect of the liquid head between the pressure source and that existing within the gauge must be considered. These gauges are generally called laboratory precision test gauges.

Classification by Accuracy

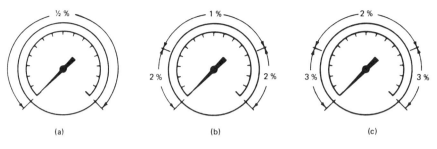

Permissible error (psi)

Span psi	Grade 2A ±1/2%	Grade A		Grade B	
		Middle 50% of span ±1%	Remainder of span ±2%	Middle 50% of span ±2%	Remainder of span ±3%
0-30	±0.15	±0.3	±0.6	±0.6	±0.9
0-100	±0.5	±1.0	±2.0	±2.0	±3.0
0-600	±3.0	±6.0	±12.0	±12.0	±18.0
0-1500	±7.5	±15.0	±30.0	±30.0	±45.0

Fig. 2.9 Examples of accuracy grades. (a) Grade 2A, (b) grade A, (c) grade B.

2.7.3 Grade 3A Gauges

Grade 3A gauges are calibrated to an accuracy of ±0.25% of span over the entire scale range and are usually offered in 4 1/2-in. diameter cases. They are generally not temperature-compensated, and use at temperatures as little as 10°F from that at which the gauge was calibrated (generally 23°C) can significantly affect the rated accuracy. These gauges are often called test gauges.

2.7.4 Grade 2A Gauges

Grade 2A gauges are calibrated to an accuracy of ±0.5% span over the entire scale range. These are the most commonly used of the higher-accuracy gauges and are generally specified by the petrochemical industry for process pressure measurements. They are often referred to as process gauges. They are usually supplied in 4 1/2-in. case size and are not temperature-compensated.

2.7.5 Grade A Gauges

Grade A gauges are calibrated to an accuracy of ±1% of span over the middle half of the scale and ±2% of span over the first and last quarters of the scale. Thus they are often referred to as two-one-two gauges. As an example of the accuracy of this grade of gauge, consider one having a span of 0–200 psi. The accuracy from 50 to 150 psi inclusive would be ±1% or ±2 psi. The accuracy from 0 to 50 psi and from 150 to 200 psi would be ±2% or ±4 psi. They may be supplied in case sizes as small as 2 1/2 in. in diameter.

2.7.6 Grade B Gauges

Grade B gauges represent the majority of those manufactured and are used for pressure measurements on equipment such as water pumps, paint sprayers, air compressors, fire extinguishers, etc. These gauges have an accuracy of ±2% of span over the middle half of the scale and ±3% of span over the first and last quarters of the scale. For example, gauges having a span of 0–600 psi would have an accuracy of ±2% or ±12 psi from 150 psi to 450 psi inclusive, and an accuracy of ±3% or ±18 psi from 0 to 150 psi and from 450 to 600 psi. These are often referred to as three-two-three gauges and are available in case sizes of 1 1/2, 2, 2 1/2, 3 1/2, and 4 1/2 in. in diameter.

2.7.7 Grade C and D Gauges

Grade C and D gauges are similar to Grade B gauges except that they are less accurate. Grade C is a four-three-four gauge, that is ±3% of span in the middle half and ±4% of span in the first and last quarters. Grade D has an accuracy of ±5% of span over the entire scale range.

2.7.8 General Comments on Accuracy

It should be obvious that higher-accuracy gauges will be more expensive and that they will require more careful handling and more frequent recalibration to be certain the accuracy is maintained.

Classification by Type of Measuring Element

Therefore, careful consideration should be given to the grade of accuracy when purchasing. Grade B, C, and D gauges are generally considered uneconomical to repair and are simply replaced if they are damaged or worn out.

2.8 CLASSIFICATION BY TYPE OF MEASURING ELEMENT

2.8.1 Types of Elements

As previously noted, the gauges converted in this handbook all use some form of elastic element that converts the measured pressure into a proportional motion. Through suitable gears and levers, this motion is transferred into a corresponding motion of the pointer to give an indication of the measured pressure. The degree of accuracy and reliability of such indications depend on the type of measuring element used. Specific requirements have resulted in the design of different types of measuring elements.

Three basic types of measuring elements are commonly used in the gauges covered in this handbook with variations of each to suit application needs:

Bourdon* elements
   C-type
   Spiral
   Helix
Bellows elements
Diaphragm elements

Figure 2.10 covers the pressure ranges normally associated with each of these elements. However, design modifications to any of them can make an exception to the generalizations given. The primary fields of application and the limitations of each type can be better understood from the descriptions that follow. Other sections of this handbook will cover specific design modifications to suit special application needs.

Bourdon is used throughout this handbook as a common noun. In the industry, the bourdon may be referred to as the bourdon tube, tube, bourdon spring, C spring, cee spring, or simply "the spring." All terms are interchangeable.

Element	Application	Minimum span (commonly supplied)	Maximum span (commonly supplied)
Bourdon	Pressure	0–12 psi	0–60,000 psi
	Vacuum	0–30 in. Hg vac	0–30 in. Hg vac
	Compound	30 in. Hg–0–15 psi	30 in. Hg vac–0–300 psi
Bellows	Pressure	0–1 in. Hg	0–100 psi
	Vacuum	0–1 in. Hg vac	0–30 in. Hg vac
	Compound	Any total span of more than 1 in. Hg	Any total span of less than 100 psi
Metallic diaphragm	Pressure	0–10 in. H$_2$O	0–10 psi
	Vacuum	0–10 in. H$_2$O	0–30 in. Hg vac
	Compound	Any total span of more than 10 in. H$_2$O	30 in. Hg. vac–0–10 psi

Fig. 2.10 Range of measuring elements.

2.8.2 Gauges Using C Type Bourdons

Gauges using a C-type bourdon (so-called because its shape resembles the letter C) as the measuring element are by far the most common. Fig. 2.11 illustrates this type of bourdon pressure element and the mechanism used to amplify the motion of the bourdon.

Before discussing the bourdon in greater detail, it will be interesting to look into its history. The name comes from its inventor, a French engineer, Eugene Bourdon. The C-shaped element (B) in Fig. 2.11 is the bourdon — the heart of the gauge.

You have probably observed how a coiled garden hose tends to straighten out when subjected to water pressure. The fact that a relatively rigid piece of metal tubing would act similarly was discovered in 1849 during the construction of a steam engine, which required a helically wound coil to condense steam. During the process of forming the coil, the tubing was flattened. To correct this,

Classification by Type of Measuring Element

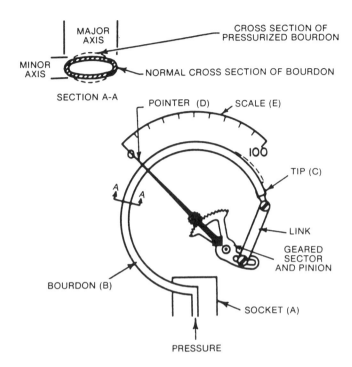

Fig. 2.11 Pressure gauge components.

the coil was plugged at one end and hydraulic pressure applied to the other end. Bourdon noticed the tube partially uncoiled as the flattened section was restored to its original circular cross section. He experimented further with this interesting action and finally invented a pressure gauge, using the end movement of a curved tube having an elliptical cross section, as a pressure-indicating device. In June 1849 Bourdon was granted a French patent.

The versatility of the bourdon pressure gauge was fully realized, as indicated by the following excerpts from U.S. Patent 9163, dated August 3, 1852:

Be it known that I, the undersigned, Eugene Bourdon, of Paris, in the Republic of France, engineer, a citizen of the Republic of France, have invented new and useful improvements in

Instruments for Measuring, Indicating and Regulating the Pressure and Temperature of Fluids. . . .

If the air be withdrawn from the interior of the tube by means of an air pump or otherwise, the curvature of the tube is increased and the index moves in the opposite direction and the instrument is then a vacuum gauge. . . .

If the air be withdrawn from the interior of the tube and it be then closed hermetically, the curvature of the tube will vary with the pressure of the atmosphere and the instrument then serves as a barometer.

The indicating pressure gauge shown in Fig. 2.11 is essentially the same as that originally designed by Eugene Bourdon. Variations and refinements, as described in this handbook, have not changed the fundamental operating principle: Pressure enters the socket A and passes into the bourdon B. The bourdon has a flattened cross section and is sealed at its tip C. Any pressure in the bourdon in excess of the external or atmospheric pressure causes the bourdon to elastically change its shape to a more circular cross section (see Section A-A, Fig. 2.11).

The result of this change in the shape of the cross section is to tend to straighten the C shape of the tube. With the socket end fixed, the straightening effect causes the tip end to move a small distance (1/16 to 1/2 in., depending on the bourdon size) proportional to the pressure. Through suitable linkage, this tip movement is used to rotate the indicating pointer D over a graduated scale E. Thus, the pressure is translated into an indication, and we have what has come to be called a dial indicating bourdon pressure gauge. Because of its inherent ruggedness, simplicity, and relatively low cost, this type of gauge has very broad applications in every field where pressure is measured.

The C-shaped bourdon gauge is manufactured in all the grades of accuracy discussed in Sec. 2.7 and can indicate vacuum, gauge pressure, or absolute pressure. The bourdon can be made from a variety of materials to suit the application, which permits its use with corrosive fluids or pulsating pressures or in unusual and severe environments. Chap. 5 describes a variety of gauge accessories that further broaden the use of bourdon gauges, as well as other types of dial indicating gauges.

As might be expected, bourdons used for indicating low pressure are made from tubing having a thinner wall than that of high-pressure bourdons. Also, low-pressure bourdons generally have a wider major axis and a smaller minor axis than high pressure bourdons

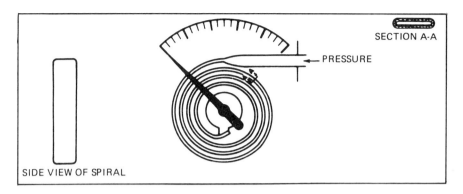

Fig. 2.12 Spiral bourdon.

(see Fig. 2.11). This combination of thin wall, large major axis, and small minor axis produces a bourdon whose cross section will deflect to provide the required rounding at a lower pressure than one having a heavy wall, short major axis, and large minor axis. There is, however, a practical limit on both wall thickness and width, so at less than 12 psi it is not practical to manufacture a bourdon that will provide sufficient force and motion at its tip end to permit the use of conventional mechanical motion multipliers. Bourdons are therefore rarely used in gauges having a span (total range) of less than 12 psi. Mechanical pressure gauges for indicating pressure lower than 12 psi use bellows or diaphragm elements as discussed below.

2.8.3 Gauges Using Spiral Type Bourdons

One variation of the C bourdon is the spiral bourdon (see Fig. 2.12). It is generally made from tubing with a more flattened cross section than a C bourdon and is designed to provide more motion at the tip and less stressing of the bourdon material, since less motion per increment of tubing length is required.

The spiral form of bourdon is made by winding the tubing in the form of a spiral having several turns, instead of the approximately 250° constant diameter arc of the C bourdon. This arrangement does not change the operating principle, but simply has the effect of producing a tip motion equal to the sum of the individual motions that would result from each part of the spiral considered as a C

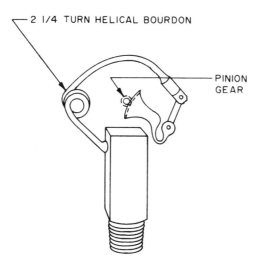

Fig. 2.13 Helical bourdon with movement.

bourdon. This type of bourdon can be made to produce sufficient motion to permit the pointer to be attached directly to the free end of the bourdon without the need to use a multiplying mechanism. Practically, it is difficult to wind sufficient tubing to obtain a pointer travel of 270°, but in pressures above 150 psi it is possible to obtain as much as 180°. Since it is not necessary to drive a multiplying mechanism, the force developed at the end of the bourdon may be very small. This style of bourdon permits the use of thin-walled, small-diameter tubing (less than 0.10 in. in diameter). Gauges using this bourdon are usually specified where low cost and small size are desired, such as for indicating the pressure in pressurized fire extinguishers. Theoretically, the element could be used for vacuum measurement, just as a C bourdon is; in practice, however, it is confined to gauge pressure.

2.8.4 Gauges Using Helical Type Bourdons

The helical bourdon element in the form shown on Fig. 2.13 is designed to be used with a multiplying movement and provides the same motion at the tip as a C bourdon. Alternatively, the bourdon can be made from a long length of small-diameter tubing so as to

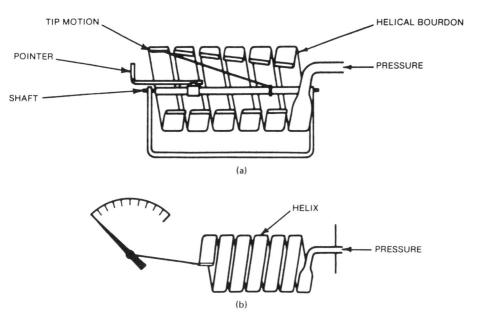

Fig. 2.14 Helical bourdon without movement. (a) Detail view of typical helix, (b) schematic.

provide sufficient travel to make a multiplying linkage unnecessary (see Fig. 2.14). With this type of bourdon, pointer travels of 270° may be attained directly, but a very long length of tubing, having up to 30 turns, is required, and a supporting shaft may be needed to fix the center of rotation.

The helical bourdon is sometimes used in recording pressure gauges, where a large motion is required to move a pen across a chart that is rotating as a function of time, thus obtaining a trace of pressure versus time. In this case, bourdons having a large major axis are used, in order to provide sufficient force at the moving end of the helix, to move the pen across the chart. Small-diameter tubing, such as that used for indication, will not develop sufficient force to drive the pen.

High-pressure bourdons are necessarily made from tubing having a thick wall, and the mechanical stress developed in the wall when pressure is applied to the bourdon is higher than that developed in low-pressure bourdons. Repeated applications of pressure result in cyclic stressing which, as with any spring, may eventually cause the bourdon to fail. Failure due to cyclic stress is called fatigue failure, and the closer the stress approaches the yield point of the material the earlier the failure will be. A bourdon utilizing small-diameter, thin-walled tubing wound into the helical configuration shown in Fig. 2.13 will be subjected to less stress than a similar C bourdon having the same tip motion in the same pressure range. Therefore, the helical bourdon will have an improved fatigue life and will withstand higher overpressure than will a comparable C bourdon.

2.8.5 Gauges Using Bellows Elements

The bellows pressure element (see Fig. 2.15) is made by hydraulically or mechanically forming thin-walled tubing or by utilizing a plating process. The figure is a simplified schematic of the manner in which a bellows pressure element can be adapted to measure pressure. However, in order to attain a reasonable fatigue life and motion that is linear with pressure, a coil spring is used to supplement the inherent spring rate of the bellows so that for a given pressure the bellows motion will be somewhat restrained (see Fig. 2.16). With this arrangement the device is called a spring-loaded bellows.

Bellows elements are most often used for the measurement of low pressure. They are available in diameters up to several inches, and therefore can provide usable motion and adequate force to drive the motion-multiplying mechanism for pressure as low as 1 in. Hg.

Bellows are available in brass, stainless steel, phosphor bronze, Monel (Huntington Alloys Inc.), and beryllium copper. Brass is most commonly used. The other materials are used in special applications, including corrosion-resistant service.

2.8.6 Gauges Using Diaphragm Elements

Metallic diaphragm elements are also used to indicate low pressure. A diaphragm plate is made by forming thin sheet material into a shallow cup having a series of concentric corrugations on the end wall (see Fig. 2.17). Two plates are joined by soldering, brazing, or welding to form a capsule, and two or more capsules may be joined to form a stack. Pressure is usually applied to the interior of the assembly. The total motion of the stack is the sum of the individual plates. Therefore, the lower the pressure to be measured or the higher the desired travel, the more plates must be used in the stack. A single capsule about 2 in. in diameter can

Classification by Type of Measuring Element   45

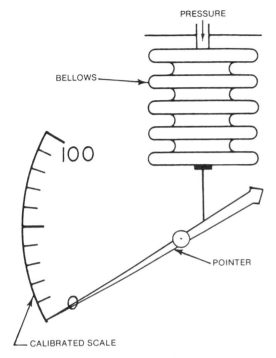

Fig. 2.15 Bellows-actuated gauge.

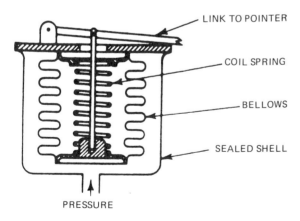

Fig. 2.16 Spring-loaded bellows.

46   Chap. 2 Fundamentals of Pressure Gauges

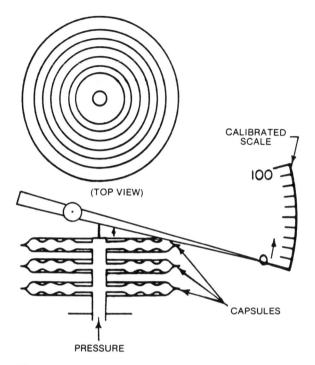

Fig. 2.17 Metallic diaphragm gauge.

provide 0.060 in. motion, which is generally sufficient to drive a
high-ratio multiplying mechanism, since diaphragm elements deliver
high force. Opposing coil springs are not generally used, as in the
case of the bellows pressure element, because the spring characteristics are provided by the diaphragms themselves.

It is possible to form the diaphragm plate so that its motion with
respect to the applied pressure will be either linear or have a controlled degree of nonlinearity. The latter is useful for such applications as altimeters, where equal increments of altitude do not result in equal increments of pressure. By properly characterizing
the individual plates of a diaphragm stack, the motion can be made
linear with altitude over a wide range.

Diaphragm elements are often used in absolute pressure gauges
because they are most suitable for low-pressure ranges where compensation for atmospheric pressure change is most significant (see

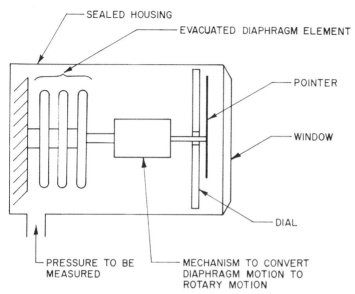

Fig. 2.18 Single diaphragm—absolute pressure gauge.

Sec. 1.5.3). There are two methods of using diaphragm elements to measure absolute pressure. One method is shown in Fig. 2.18. The diaphragm element is evacuated, hermetically sealed, and placed inside a closed chamber. The pressure to be measured is admitted to the chamber and surrounds the diaphragm element. Changes in the measured pressure will cause the element to deflect, but changes in atmospheric pressure are excluded and therefore have no effect, and the gauge may be calibrated in terms of absolute pressure. This is the method used in aneroid barometers, wherein the input pressure is atmospheric pressure.

A second method is illustrated in Fig. 2.19. This method uses two diaphragm elements arranged so as to oppose each other. One element (A) is evacuated and sealed. The pressure to be measured is applied to the interior of the other element (B). When pressure is applied to element B, it expands and partially compresses element A. The linear motion of the rigid shaft between the two elements is converted to rotary motion of the pointer. An increase or a decrease in atmospheric pressure will act equally on both elements so that no motion will result, and therefore the gauge may be calibrated in terms of absolute pressure. The

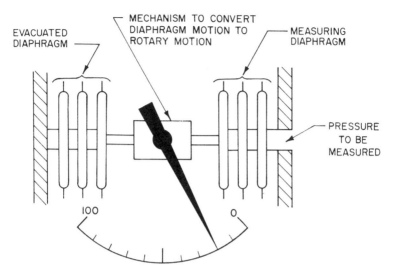

Fig. 2.19 Opposing diaphragm—absolute pressure gauge.

evacuated capsule provides a fixed reference point at zero absolute pressure.

The reason for evacuating the measuring element shown in Fig. 2.18 and element A (referred to as the reference element) of Fig. 2.19 is to avoid a change in indication with temperature. If the element were simply sealed, leaving air inside, then as the temperature surrounding it (ambient temperature) increased, the air pressure inside the element would increase and impart some motion to the mechanism, which would appear as a decrease in the absolute pressure being measured. By eliminating the air from the element, this error is avoided.

Both of the arrangements just described can be used to indicate the difference between two pressures (differential pressure). In this case, the diaphragm element is not evacuated and sealed but is connected to one of the two pressures to be measured. The other pressure is connected as shown in the diagrams, that is, either to the interior of the case (Fig. 2.18) or element B (Fig. 2.19). The difference between the two pressures causes a resultant motion, and this gives an indication of the differential pressure.

Diaphragm elements are commonly made of phosphor bronze, because of the excellent spring characteristics of this alloy. Other materials include beryllium copper, stainless steel, Ni-Span C (International Nickel Co.), and Monel.

For the measurement of pressure less than 10 psi, diaphragm gauges are generally superior in performance to bourdon gauges. Because the motion of the diaphragm may be quite small, the gauge requires a high-quality, low-friction movement to translate the diaphragm motion into motion of the indicating pointer. On the other hand, because of its relatively large effective area, the diaphragm develops many more times the pointer torque (turning force on the pointer shaft) than a bourdon does at the same pressure, thus giving better repeatability and higher stability.

Typical applications for diaphragm gauges include: absolute pressure gauges, draft gauges, pneumatic receiver gauges, liquid level gauges, low-pressure differential pressure gauges, and gauges for indicating the pressure drop across air filters. When gauges constructed in the manner described above are used for the measurement of differential pressure, the static pressure must be considered. That is, the gauge may be suitable for indicating the difference between 10 psi and 12 psi, but not suitable for indicating the difference between 100 psi and 102 psi. In other words, the 100 psi static pressure exceeds the ability of the case of Fig. 2.18, or the elements of Fig. 2.19, to contain it. Also, the consequences must be considered of inadvertently applying maximum pressure to one side of the element and zero pressure to the other side.

3
Gauge Components

3.1 BASIC COMPONENTS

Many of the parts that make up a specific model of a pressure gauge are the same or quite similar — regardless of whether the measuring element is a bourdon, metallic diaphragm, or spring-loaded bellows. This chapter deals mainly with the bourdon-actuated gauge and covers its basic components, their primary functions, and the numerous terms used to identify them. Specific model variations and component variations are covered at appropriate points in the text.

3.1.1 Nomenclature of Basic Components

The typical bourdon gauge (Fig. 3.1) has the following basic components:

1. Measuring element — the closed chamber within which the pressure medium is confined. In this case it is the bourdon; however, it may also be a diaphragm or bellows. It must be connected to the socket, either directly or through capillary.
2. Socket or pressure connection — the fitting that extends out of the case and to which the measured pressure is connected. It is sometimes called the stem.
3. Tip — the component fastened to the free end of the bourdon to provide a closure and allow for the attachment of a link.

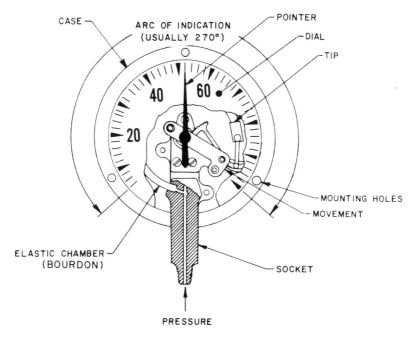

Fig. 3.1 Bourdon gauge.

4. Movement — the mechanism that amplifies the motion of the measuring element to provide pointer rotation.
5. Dial — the plate on which the graduated scale is displayed.
6. Pointer — the member that travels over the scale to indicate the pressure applied to the measuring element.
7. Case — the protective housing for the gauge parts.
8. Ring — the retaining ring that secures the window to the case.
9. Window — the transparent cover that is secured in front of the dial by the ring.

3.2 SOCKET

3.2.1 Location of Pressure Connection

The socket functions as a means of connecting the gauge to the pressure source. It usually provides support for all other gauge

Socket

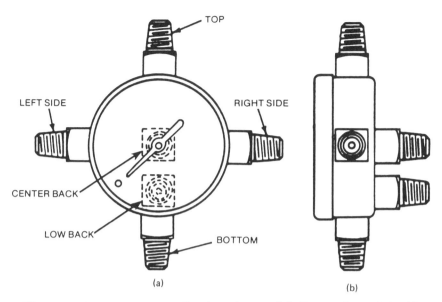

Fig. 3.2 Pressure connection locations. (a) Front view, (b) side view.

components. The threaded end is usually an external (male) American Standard taper pipe thread (abbreviated NPT). The socket can extend from the gauge case (Fig. 3.2) in various positions to suit the application.

1. Center-back connection -- the socket extends from the center of the case perpendicular to the case back.
2. Low-back connection − the socket extends from the back of the case but below the center, usually on the vertical centerline.
3. Low or top connection − the socket extends radially from the side of the case at the bottom or top, as shown. The low connection is most common and is also referred to as bottom connection.
4. Right or left connection − the socket extends radially from the side of the case, at the right or left side.

Combined with this terminology are such expressions as "low male" connection (abbreviated LM), "center-back male" connection

(CBM), or "low-back male" connection (LBM). These terms specify a male pipe thread, as well as the location of the connection.

3.2.2 Connection Means

Although the most common thread size is 1/4-in. NPT male, the application may call for a smaller fitting, such as 1/8-in. NPT male, or a larger fitting, such as 1/2-in. NPT male, and sockets can be made in these sizes. However, the socket should be in proportion to the physical size of the gauge, and this somewhat restricts the size of connection. For example, gauges with a 1 1/2-in. dial size usually have a 1/8-in. NPT male connection; larger threaded sizes would be far too bulky for the size of the gauge. On the other hand, as the size of the gauge and magnitude of the pressure increases, a 1/4-in. NPT may not be adequate for direct stem support of the gauge. Therefore, a 1/2-in. NPT male connection is generally standard for gauges with dial sizes of 4 1/2 in. or greater, particularly when stem-mounted. In addition to the standard pipe threads, a variety of other connections is generally available on special order. These include:

Female pipe threads (NPT)
Alligator or hose fittings
Flare fittings
Dry-seal threads (NPTF)
Metric straight or tapered pipe threads
Special high-pressure threads

3.2.3 Materials

Sockets are manufactured in a variety of materials to suit the application. Materials commonly used include:

Brass
Alloy steel
Stainless steel
Monel

The socket is generally made from bar stock. Forgings and castings are sometimes used, but their high cost limits their usage.

3.2.4 Bar Stock Sockets

Bar stock sockets are so named because they are made from a length of extruded solid bar with the desired cross section. For example, bar stock with a square cross section is generally used for low-connected gauges, while a specially shaped cross section is used for center-back-connected gauges. The bar stock socket is economical to make in large quantities and is widely used on gauges in all classes of accuracy and in all the common materials. A portion of the square bar is made to protrude from the case and provides a wrench flat for installation.

3.2.5 Forged or Cast Sockets

Sockets made from forgings usually provide bosses for mounting the movement, dial, and case, thus providing a rugged unitary structure for all the gauge components. However, they are very expensive, partly because they require more material than bar stock sockets do. This can be a significant factor, especially if the socket is made of stainless steel or Monel. Sand-cast sockets generally exhibit unacceptable porosity and are rarely used. Forged sockets are more difficult to adapt to automatic machining. For these reasons forged sockets have been largely supplanted by bar stock sockets.

3.3 MEASURING ELEMENT

As covered in Sec. 2.8.1, three basic types of elastic elements are employed in the gauges covered by this manual. Each is considered in further detail as to the materials and variations in fundamental designs.

3.4 BOURDONS

The design of a bourdon for a specific application involves a number of considerations with respect to factors such as material, coiling length, wall thickness, major and minor axis size, shape of cross section, pressure range, and tip travel. A complete understanding of all these factors is not essential to the user. However, some discussion will provide an appreciation of the uses and limitations of bourdons and thereby assist the user in making a proper selection.

3.4.1 Design Factors

In general, there is no one bourdon that is best under all conditions. For example, type 316 stainless steel offers high resistance to corrosion for a wide variety of chemicals. However, it is expensive and cannot be fabricated to attain the high tensile strength that a heat-treatable alloy such as beryllium copper can attain. Therefore, if high overpressure and low hysteresis are requirements, then a compromise must be made with respect to corrosion resistance. On the other hand, if corrosion resistance is paramount, then either some hysteresis must be tolerated or the bourdon must be designed to produce a smaller tip travel, have a longer coiling length, or be able by some other means to keep the operating stress level well below the yield point of the material. Class B commercial gauges are used in very large quantities and cost becomes an important consideration. Therefore selection of the size of the bourdon and the material must enable the manufacturer to produce a gauge that is capable of meeting the user's performance requirements, but does not result in the user paying for unnecessary features. Let's look at the bourdon in greater detail.

3.4.2 Explanation for Motion at Tip of Bourdon

The motion at the free end of the bourdon (the end opposite from that joined to the socket) that occurs when pressure is applied to the interior of the bourdon is called the tip travel. The tip travel results from the stresses set up within the walls of the tube when the flattened cross section of the bourdon is forced by the applied pressure into a more nearly round section. Due to the fact that the bourdon is formed into an arc, this elastic deformation of the cross section will tend to create a compressive stress in the wall of the cross section which is nearest to the center of the coiling radius (inner wall) and a tensile stress in the wall of the cross section which is furtherest from the center of the coiling radius (outer wall). A study of Fig. 3.3a will show why this occurs. These stresses, added algebraically to the tensile stress in both walls created by the internal pressure, result in an unbalance such that the tensile stress in the outer wall is higher than the tensile stress in the inner wall. Uncoiling of the bourdon (i.e., an increase in the coiling radius) tends to create a tensile stress in the inner wall and a compressive stress in the outer wall, which is the opposite condition caused by the change in cross section and which will relieve the unbalance created by the change in cross section. Motion at the free end of the bourdon will therefore occur in the approximate direction shown in Fig. 3.3, until the opposing stresses come into equilibrium.

When this principle is understood, it becomes clear why a thinner wall, a larger major axis, and a smaller minor axis all tend to

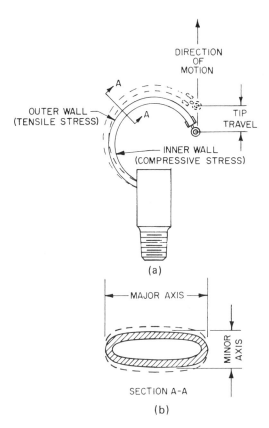

Fig. 3.3 Motion of bourdon.

increase the tip travel for a given pressure whereas the opposite conditions will tend to decrease the tip travel.

The full-scale tip travel is quite small (less than 1/4 in., even for a 6-in. gauge). This motion must therefore be amplified greatly to provide full-scale travel of the pointer across the indicating scale (generally a 270° arc) or approximately 10 in. of scale length for a 4 1/2-in. gauge. The gauge "movement" (link, sector, and pinion) provides this amplification of tip travel by its lever and gear action.

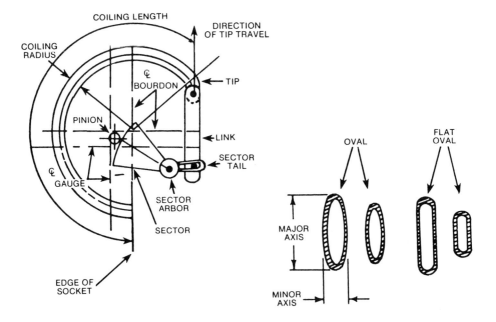

Fig. 3.4 Basic bourdon gauge layout.

3.4.3 Geometry of Layout

The basic layout of a bourdon pressure gauge is shown in Fig. 3.4. Using this type of diagram, the gauge designer will determine the proper length of the sector tail and the gear ratio between the pinion and the segment that will be required to produce a pointer rotation of 270° for the nominal tip travel of the bourdon being used. Conversely, he may determine the tip travel required to operate a specific movement. The tip travel of a bourdon is generally quite linear with pressure, that is, the amount of tip travel at say 25%, 50%, and 75% of the total applied pressure is equal to 25%, 50%, and 75% of the total travel. The gauge designer must select and position the movement so as to give equal increments of rotation for equal increments of bourdon tip travel. In this way the pointer will move through the same angular travel for the same pressure change over the entire scale, in which case the pointer motion is said to be linear. For gauges in the accuracy grade of 2A and higher, the geometry of the layout becomes critical with respect to achieving

"C" Bourdon as shown in Fig. 2.11			Helical bourdon as shown in Fig. 2.13	
High pressure	Low pressure	Coiling radius	High pressure	Coiling radius
0.080	0.090	0.63	0.090	0.16
0.095	0.115	0.81	0.115	0.20
0.095	0.200	1.63	0.200	0.88

Fig. 3.5 Typical tip travel of bourdons.

the linearity necessary for high accuracy. Such high-accuracy gauges usually have some means to vary the geometric relationships and thereby adjust for linearity. Chapter 7, "Gauge Maintenance and Calibration," discusses and illustrates how linearity adjustments are made.

3.4.4 Tip Travel

For the purpose of providing a general guide, the tip travel and coiling radius for typical bourdons are given in Fig. 3.5. This chart shows that the nominal tip travel for a series of C bourdons, covering ranges from 0-15 psi through 0-10,000 psi, will vary widely with the range of the gauge. The reason for this is that it is difficult to overcome the friction inherent in the amplifying movement, utilizing the low force provided by the thin walls and small minor axes associated with low pressure bourdons. Smoother pointer motion will be obtained by having a high tip travel and using a movement having low amplification. However, such a high tip travel will produce excessive stress levels in high-pressure bourdons, resulting in early fatigue failure. The heavier walls and larger minor axes associated with high-pressure bourdons provide a higher force at the tip so that it is possible to utilize a high-amplification movement and keep the tip travel small. Theoretically there is an optimum tip travel (and corresponding movement) for each pressure range. However, providing such a large number of different movements complicates the manufacture and maintenance of the gauge, and the small gain in performance that would result is not justified. Generally the tip travel is standardized at two or three levels, each level corresponding to a movement of given magnification ratio to cover the entire series from 0-15 to 0-10,000 psi. Movement gear ratios will vary accordingly, and might be as low as 8:1 or as high as 20:1.

Fig. 3.5 also shows that the tip travel for high pressure helical bourdons of the type shown in Fig. 2.13 is the same as for low-pressure C bourdons. Helical bourdons of this type do not have the high force of a C bourdon in the same pressure span. Therefore, as with low-pressure C bourdons, low ratio movements and high tip travels are needed to obtain smooth pointer motion. In spite of the high tip travel, helical bourdons have better fatigue life compared to C bourdons because of the low stress levels that can be obtained, and the high tip travel affords the opportunity to have a single ratio movement for the whole series of gauges from 0-15 psi through 0-15000 psi.

As previously mentioned, the cross-sectional shape of a bourdon is a factor in determining tip travel. The shape of the cross section may be oval or "flat oval" as shown in Fig. 3.3. The flat oval is the most commonly used as it requires only flattening rolls rather than the shaped rolls necessary to produce the oval cross section.

3.4.5 Overpressure Limit

Overpressure limit is another important design consideration in the proper selection of the bourdon. It is defined as the pressure above the full-scale value of the gauge that the gauge will tolerate without a permanent set due to exceeding the elastic limit of the bourdon material. It is often expressed as a percentage of full scale (for example, a 100 psi gauge that can withstand a maximum of 150 psi is said to withstand a 50% overpressure). However, specifying overpressure in terms of percentage can become confusing. For instance, "150% overpressure" on a 0-100 psi gauge might be interpreted as either 150 psi or 250 psi maximum applied pressure. It is therefore best to state the maximum pressure to be applied to the gauge. For example, a gauge with a 0-100 psi range that must withstand a maximum pressure of 150 psi should be specified as "Range 0-100 psi; maximum applied pressure 150 psi." The ability of a bourdon to withstand overpressure depends on the material from which it is made and its geometry. Because they operate at stresses closer to the yield point, high-pressure bourdons will not withstand the same degree of overpressure as low pressure bourdons will, and also will have a lower fatigue life. The fatigue life may be defined as the number of times the bourdon can be pulsed over some portion of its span before it fails.

Gauge manufacturers have data that serve as a general guide to their engineers as to the permissible overpressures for bourdons in their standard line. No broad generalization can be made that applies to all types of gauges. To determine the overpressure allowable, the manufacturer must consider the type and size of gauge, the required accuracy, the spring material, the pressure range, and whether the unit is equipped with a "maximum stop" (described under Sec. 3.12.6.5), which restricts the bourdon to the useful limit (full-scale value) of its travel.

To stabilize a bourdon so that it will retain its calibration when subjected to the specified overpressure, the manufacturer must subject the gauge to this value several times prior to calibration of the gauge. Therefore, whenever the application requires the gauge to withstand overpressure, it should be specified in the purchase description so that the gauge manufacturer will impose this overpressure on the gauge prior to calibration. Unless it is known that the gauge has been specifically manufactured to withstand overpressure, it is best not to pressurize the gauge beyond full scale.

3.4.6 Materials for Bourdons

Bourdons can be made from any alloy that exhibits satisfactory mechanical properties. They are commonly divided into nonferrous or ferrous groups, that is, whether they contain no significant amount of iron (Fig. 3.6), such as admiralty brass, phosphor bronze, beryllium copper, Monel, and K Monel (Huntington Alloys Inc.), or do contain a significant amount of iron (Fig. 3.7), such as alloy steels.

1. Nonferrous bourdons. The materials listed in Fig. 3.6 are by no means the only types of nonferrous material from which bourdons can be made, but rather a selected list of the materials most commonly used by gauge manufacturers because they exhibit the most desirable properties. Admiralty brass is generally used in inexpensive, low-pressure, grade B accuracy gauges. Phosphor bronze is generally used in higher-quality grade A and AA gauges, although some manufacturers offer this material in grade B gauges as well, because of its ability to withstand more overpressure without taking a permanent set. Observe the comparison of yield strength (Fig. 3.8), which is related to the ability of a bourdon to withstand overpressure. Beryllium copper is one of the few copper alloys that is hardenable by heat treatment. The other copper base alloys obtain their hardness or temper from the cold working imparted to them during the tube drawing process. Beryllium copper may be formed in an annealed or slightly cold-worked state and then heat-treated afterward to obtain the desired tensile strength. These properties make beryllium copper ideal for high-pressure applications (over 1000 psi) or for lower-pressure applications where extreme overpressures are to be encountered.

Material	Cu (%)	Zn (%)	Sn (%)	P (%)	Be (%)	Fe (%)	Mn (%)	Ni (%)	Other
Admiralty Brass UNS C44300	70/73	Balance	0.90/ 1.20						
Phosphor Bronze UNS C51000	94 min	0.30 max	3.50/ 5.80	0.03/ 0.35		0.10 max			Pb = 0.05 max
Beryllium Copper UNS C17200	Balance				1.80/ 2.00				Co + Ni + Fe = 0.60 max Co + Ni = 0.20 min
Monel UNS N04400	Balance					2.50 max	2.00 max	63/70	C = 0.30 max Si = 0.50 max
K-Monel UNS N05500	Balance					2.00 max	1.50 max	63/70	C = 0.25 max Si = 1.00 max Al = 2.00/4.00

Fig. 3.6 Chemical composition – nonferrous bourdons.

Material	C (%)	Cr (%)	Ni (%)	Mn (%)	Si (%)	Mo (%)	Fe (%)	Other
Type 316 St. Steel UNS S31600	0.08 max	16/18	10/14	2.00 max	1.00 max	2.0/3.0	Balance	
Type 403 St. Steel UNS S40300	0.15 max	11.5/ 13.0		1.00 max	0.50 max		Balance	P = 0.04 max
4130 Alloy Steel UNS G41300	0.28/ 0.33	0.80/ 1.10		0.40/ 0.60	0.15/ 0.30	0.15/ 0.25	Balance	
Ni-Span "C" UNS N09902	0.06 max	4.90/ 5.75	41/43	0.80 max	1.00 max		Balance	Al = 0.30/0.80 Ti = 2.10/2.75

Fig. 3.7 Chemical composition — ferrous bourdons.

Material	Tensile strength (psi x 1000)	Yield strength (psi x 1000)	Elastic modulus (psi x 10^6)
Admiralty Brass UNS C44300	53	22 (annealed)	15
Phosphor Bronze UNS C51000	68	55 (1/2 hard)	16
Beryllium Copper UNS C17200	190/165	125/100	19
Monel UNS N04400	84/78	41/33 (annealed)	26
K-Monel UNS N05500	140	100	26

Note: Values are given in the cold-worked or heat-treated condition generally used for bourdons.

Fig. 3.8 Physical properties — nonferrous materials.

Monel is a nickel base corrosion-resistant alloy for applications under 1000 psi. It obtains its spring properties through cold work. K-Monel is also a nickel base alloy; it can be heat-treated to a higher hardness than Monel can and may be used for high-pressure applications. It has reasonably good spring properties, although not quite as good as those of beryllium copper. However, it does have the advantage of corrosion resistance and is usually selected for a particular application when other materials would be attacked by the measured fluid. Spring properties are obtained by a lengthy heat-treating cycle, which adds to the cost of the gauge.

2. Ferrous bourdons. The chemical compositions of the more common ferrous materials are listed in Fig. 3.7. The physical properties are listed in Fig. 3.9, and, again, the yield strength is a good indication of the ability of a bourdon to withstand overpressure. Type 316 stainless steel is highly resistant to most chemicals. Like phosphor bronze, this material obtains its spring properties entirely from cold work and cannot be heat-treated to an increased degree of hardness. Therefore, it is usually used in applications below 10,000 psi. When it is necessary to use the

Material	Tensile strength (psi x 1000)	Yield strength (psi x 1000)	Elastic modulus (psi x 10^6)
Type 316 St. Steel UNS S31600	115/85	100/45 (cold worked)	28
Type 403 St. Steel UNS S40300	120/70	90/40	29
4130 Alloy Steel UNS G41300	256/140	220/97	30
Ni-Span "C" UNS N09902	193/131	173/126	24/47

Note: Values are given in the cold-worked or heat-treated condition generally used for bourdons.

Fig. 3.9 Physical properties — ferrous materials.

material for higher pressures, increased accuracy tolerances are normally required. Type 403 stainless steel is not as corrosion-resistant as type 316 stainless steel but can be used for pressures above 10,000 psi. Unlike type 316 stainless steel, this material is heat-treatable to a high hardness, with suitable elongation. Type 4130 is an alloy steel that does not have the corrosion resistance of the stainless steel alloys. It obtains its spring properties through heat treatment and is satisfactory for pressures up to 0-50,000 psi where corrosive media are not encountered. Ni-Span C is a heat-treatable, high-nickel alloy that is used for its superior temperature behavior. Bourdons fabricated from most other metals undergo a change in spring rate with a change in ambient temperature and thereby introduce an error in indication when they are used at a temperature other than that at which they were calibrated. A Ni-Span C bourdon will maintain its spring rate since Ni-Span has a zero thermoelastic coefficient over the temperature range of -50 to +150 F (see Sec. 6.3 for further discussion of temperature effect). Ni-Span C possesses elastic spring properties on a par with beryllium copper, but is very expensive and requires a long heat treatment in a vacuum furnace, which contributes materially to the cost of the gauge.

66   Chap. 3 Gauge Components

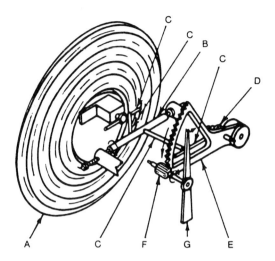

Fig. 3.10 Diaphragm gauge. A, diaphragm; B, rocker arm; C, actuating pins; D, arbor; E, sector; F, pinion; G, pointer.

3.5 METALLIC DIAPHRAGMS

A diaphragm element with one or several capsules forms a compact assembly that requires a specially designed lever system to transmit its small motion into rotary motion of the indicating pointer. Fig. 3.10 shows a typical movement wherein the diaphragm operates a rocker arm which in turn rotates a geared segment and pinion.

3.5.1 Design -- Fabrication

The design and fabrication of diaphragm pressure elements make up a special field. There is some literature on the subject that is very useful, particularly for the design of diaphragms having motion which is linear with pressure. For characterized diaphragms, that is, those having motion linear with altitude, flow, airspeed, etc., the design is essentially a matter of cut-and-try. The several factors that enter into the relationship of motion versus pressure are the type of material, thickness of material, depth and number of

Metallic Diaphragms

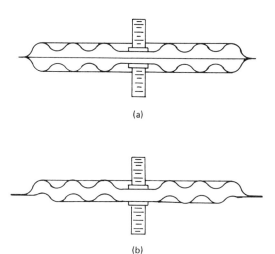

Fig. 3.11 Diaphragm capsule. (a) Conventional, (b) nested.

corrugations, shape of corrugations, cupping height, flatness of plates (i.e., crowned or dished), amount of travel, and method of joining plates.

Diaphragm elements are usually made as illustrated in Fig. 3.11a, wherein two identical plates are used to make a capsule. An alternative design is illustrated in Fig. 3.11b, wherein the two plates forming the capsule are nested. This construction takes up less space than the conventional construction does, which is often advantageous where considerable motion is required in a small space.

Another advantage of the nested construction is that a relatively high external pressure can be applied to the element without damage, since the lower plate can be formed to nest into the upper plate, where they will support each other. This construction is useful for low-range, absolute pressure gauges (for example, 0—1 psia) of the type shown in Fig. 2.18, where the element is subjected to full atmospheric pressure (0—15 psia) when the pressure connection of the gauge is open to atmospheric pressure. The disadvantage of this construction is that two different sets of plates must be made, requiring two different forming dies.

3.5.2 Materials for Diaphragms

As noted in Sec. 2.8.6, diaphragm elements may be made of beryllium copper, phosphor bronze, type 316 stainless steel, Ni-Span C, and Monel. Phosphor bronze is the most commonly used material. Beryllium copper is used where overpressure and/or fatigue life make phosphor bronze unsuitable. Type 316 stainless steel and Monel are used for corrosion-resistant applications, and Ni-Span C where its constant thermal modulus is required. Other material having special corrosion-resistant properties, such as tantalum, may also be used, especially in chemical seals (see Sec. 5.2).

Diaphragm plates are formed from sheet material in specially constructed forming dies that can be adjusted to vary the depth of the corrugation and the height of the cup. The dies may also be capable of being set up to form a dished or crowned plate in order to obtain nonlinear functions. Ni-Span C and stainless steel diaphragm plates are silver brazed or welded at the lips. Beryllium copper and phosphor bronze plates are usually soft soldered, although they may be welded by using electron beam or laser welding techniques.

3.6 FORMED METALLIC BELLOWS

Formed metallic bellows of the type shown in Fig. 3.12 are often used in compression; that is, they are mounted in a container and pressure is applied around the bellows so as to compress it. If pressure is applied to the interior of the bellows so as to extend it, the useful motion per convolution, before exceeding the elastic limit of the material, is greatly reduced.

3.6.1 Design and Fabrication

Metallic bellows are necessarily made of thin-walled material, and therefore have a low spring rate (increment of force per increment of motion). This is an advantage when measuring very low pressure. However, when measuring higher pressures, it is usually necessary to employ an opposing coil spring in the manner shown in Fig. 3.12. The spring rate of the assembly thus becomes the combined spring rate of the bellows and the coil spring. By properly selecting the coil spring, the desired overall spring rate may be obtained. In this way the same bellows may be used for a variety of pressure ranges by simply changing the opposing coil spring.

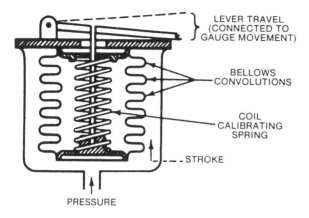

Fig. 3.12 Bellows with auxiliary spring.

The amount of motion that each convolution of the bellows can contribute is small. For example, a bellows having an outside diameter of 1 1/2 in. and a wall thickness of 0.0045 in. has a maximum permissible travel of less than 0.050 in. per convolution. If the wall thickness is increased to 0.0095 in., the permissible travel per convolution is reduced to 0.032 in. For good fatigue life and low hysteresis it is best to use only about one-half of these values. Therefore, to obtain a reasonable amount of travel to operate the indicating pointer generally requires a bellows having several convolutions. In the example given above, the height per convolution is approximately 1/8 in., so that if a bellows with four convolutions is chosen the active length will be 1/2 in. Adding the space needed for end fittings and the outside container results in an element that is much longer in the direction of travel than a diaphragm element, and much larger in internal volume.

Because of the length of a bellows element, it is usual practice to mount it so that the direction of the motion is similar to that obtained with bourdons. In this manner a movement similar to that of bourdons can be employed rather than the rocker-arm-type movement illustrated in Fig. 3.10. A lever system, as shown in Fig. 3.12, can transmit the motion to the movement. It is probable that this combination of parts will require more space than a diaphragm and rocker arm movement will, so bellows are not generally used in the smaller gauges or in very low-pressure applications.

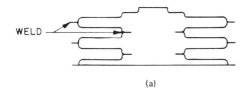

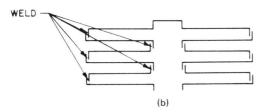

Fig. 3.13 Welded bellows. (a) Stepped, (b) cupped.

3.6.2 Materials for Bellows

Bellows elements come in various configurations, sizes, and materials as discussed in Chap. 2. Their selection for corrosion resistance, measurement of liquids with entrained solids, and other application factors involves essentially the same considerations as for gauges with bourdon elements. Because of the severe forming required, formed bellows are not available in the wide variety of materials that may be used for diaphragms.

3.7 WELDED BELLOWS

Pressure elements are commercially available that combine some of the characteristics of both diaphragms and formed bellows. These elements are constructed of individually formed plates shaped somewhat like a diaphragm. These plates are welded at their outer and inner periphery as shown in Fig. 3.13. Various shapes are used, the two illustrated being the most common. This type of bellows may be used in compression or extension or both, in the same manner as a diaphragm assembly. The large internal diameter permits welding at both the outer and inner periphery.

3.8 NONMETALLIC PRESSURE ELEMENTS

Nonmetallic pressure elements are generally made of a elastomer-coated fabric that has essentially a zero spring rate and hence is often termed a "slack" diaphragm. Slack diaphragms must be used in conjunction with some type of metallic loading spring to provide the required relationship between motion and pressure. They may be simply flat discs or discs with one or more formed corrugations, or, where a long stroke is required, they may be made having a rolling convolution operating between a piston and cylinder (see Fig. 3.14).

3.9 CHOOSING THE MEASURING ELEMENT

It will be apparent from the foregoing discussion that a variety of pressure elements is available to the gauge designer. The choice depends primarily on the required performance and secondarily on the cost. Where either of two elements will meet the performance requirements, including accuracy and reliability, then the least costly would be the usual choice. For pressure measurement above 100 psi, some form of the bourdon is normally selected (i.e., C shape, helical, or spiral). Bourdons may be used to give a 270° pointer rotation for ranges as low as 12 psi and are less costly than the other types of elements. For lower pressure, the diaphragm or bellows element is used, and for very low pressure a slack diaphragm is required. Metallic diaphragms can be formed of materials that do not lend themselves to metallic bellows. One important example is the use of Ni-Span C for diaphragm elements. Also, diaphragm elements can be made with a motion that is nonlinear with pressure (i.e., linear with altitude, air speed, flow), whereas formed bellows almost always have a linear-with-pressure characteristic. Bellows elements generally exhibit a much larger volume change with pressure than do diaphragm elements or bourdons. This may be important where a high response rate is desired or where the gauge is used in conjunction with a diaphragm seal (see Chap. 5, "Gauge Accessories," for further discussion on diaphragm seals).

3.10 TIPS

The tip is the component fastened to the moving end or free end of the bourdon to (1) close the end so that pressure can be applied

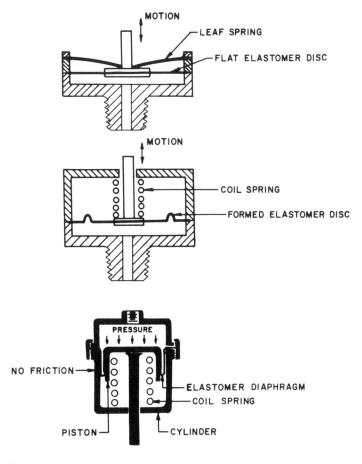

Fig. 3.14 Nonmetallic diaphragms.

inside the bourdon and (2) provide a means for attaching the link to the bourdon so that the bourdon motion can be utilized to drive the movement.

3.10.1 Design

Tips are usually made of the same material as the socket on the assumption that the two parts should have similar resistance to corrosion. All tips, with the possible exception of those used on

Tips

low-pressure gauges (under 100 psi), are designed to be fitted over the bourdon (female tip) rather than project inside the tubing (male tip). The reason for this is that the strength of the joint between the bourdon and the male tip depends completely on the bonding material, since no mechanical joining means, such as crimping, staking, or riveting, is used. When the joint is made with soft solder, as is most common, the pressure applied to the bourdon will tend to peel the thin wall of the bourdon away from a male tip. The shear strength of the soft solder will resist the action of the pressure to drive the male tip out of the bourdon, but soft solder has little resistance to peel, wherein the solder at the immediate interface between the two parts is subjected to a tensile stress. Brazing or welding the male tip greatly reduces the possibility of peeling, but it is not always feasible to use these methods.

Female tips are not subjected to peeling forces, and the shear strength of the entire soldered area must be exceeded to push the tip off the bourdon. Further, it is less likely that solder flux will be trapped inside the bourdon when attaching female tips, and it is easier to apply the flux to the outside of the bourdon rather than the inside. Therefore, as previously noted, this style is much preferred.

3.10.2 Fabrication

The female tip may be made as a channel, open at the ends, or it may be made as a cup completely surrounding the bourdon. Gauges having pressure ranges above 1000 psi usually use the cup type for added strength (see Fig. 3.15).

Several means of fabrication are available to manufacture the tips. They may be stamped from sheet, forged, cast, or milled from bar. However, stamping is the most common, because it permits a variety of materials using the same stamping die, and because it is generally less expensive than the other methods. The tab to which the link is attached may be an integral part of the tip or it may be added by welding (Fig. 3.15).

3.10.3 Alternate Method of Closing Bourdon

It is sometimes possible, particularly when ferrous alloys are used for the bourdon, to close the bourdon by simply flattening the end shut and welding across the joint. The link take-off is welded on the flat of the bourdon. If this method is used, care must be taken not to crack the tubing where it is severely formed. This is especially true on bourdons for high-pressure ranges having heavy walls.

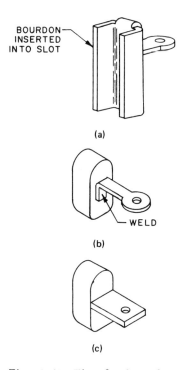

Fig. 3.15 Tips for bourdons. (a) One-piece stamping, (b) two-piece stamped cup, (c) forged cup.

3.11 ASSEMBLY OF SOCKET, BOURDON, AND TIP

The method used to assemble the socket, bourdon, and tip must be given careful consideration. Factors that must be taken into account include:

Materials to be joined
Corrosion resistance of joint
Annealing temperature of bourdon material
Strength of joint with respect to the pressure to be measured
Maximum temperature to which the assembly will be exposed
Manufacturability of joint

Assembly of Socket, Bourdon, and Tip 75

Various methods can be considered, including soft solder, high-temperature hard solder, silver brazing, nickel brazing, and welding.

3.11.1 Soft Soldering

Soft soldering, using a 50% tin, 50% lead alloy, is the most common joining method, and the great majority of gauges, particularly the class B commercial gauges, use this method. It is most suitable for use with copper alloys, i.e., brass and phosphor bronze. However, gauges for ammonia service using steel sockets, bourdons, and tips are frequently soft soldered. Soft solder joints are readily made on automatic soldering equipment using either wire solder or solder paste (small particles of solder alloy mixed with flux so as to form a paste). The strength of the joint is adequate for gauges having full-scale range up to 2000 psi, and the temperature to which the parts must be raised to flow the solder (425°F) does not anneal the bourdon. Soft solder joints should not be continuously exposed to ambient temperature above 120°F and not exposed at all to temperatures above 160°F. At these elevated temperatures the shear strength of the soft solder is reduced, and, particularly with gauges in the higher-pressure ranges, it is possible to eventually separate the bourdon from the socket or the tip from the bourdon.

Two machining methods are used to form a cavity in the socket to receive and locate the bourdon. For high-pressure gauges (over 600 psi) the socket cavity is spline milled; that is, the cavity does not extend to the outer edge of the socket, and the bourdon is completely surrounded by the socket material as shown in Fig. 3.16. The cavity is dimensioned so as to provide a small clearance around the bourdon. The depth of the cavity is on the order of 1/8 in. The second method is to mill a slot completely through the socket as shown in Fig. 3.17. This method does not give support to the bourdon at the edges but does lend itself to higher production rates compared to the spline milling, and it has been found to be quite satisfactory for low-pressure gauges.

3.11.2 Silver Brazing

For higher-pressure ranges and for bourdons having a larger cross section (such as those used in process gauges and test gauges) it is better to use a brazing alloy as the joining means. Silver brazing alloys are commercially available in a variety of compositions. One commonly used alloy consists of 45% silver, 15% copper, 16% zinc, and 24% cadmium, and has a flow point of 1145°F. Silver brazing

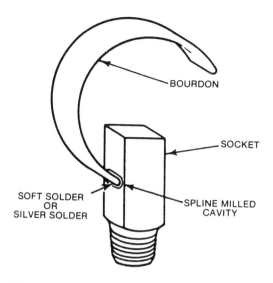

Fig. 3.16 Method of attaching bourdon — spline milled.

alloys are commonly referred to as silver solder, which is not strictly correct since solders are considered to have a melting point below 800°F. Silver solder has good corrosion resistance and may be used to join both copper and ferrous alloys. It is preferred for joining beryllium copper. Due to the high temperature required to flow silver solder, the soldering operation must be done carefully so as not to anneal the bourdon excessively at or near the solder joint. Silver solder is more expensive than the tin-lead solders and therefore is used only when required.

When silver solder is used as the joining method, the cavity is spline milled (see Fig. 3.16). However, because of the higher strength of silver solder compared to soft solder, the depth of the cavity may be reduced to as little as 1/16 in.

3.11.3 Welding

When a stainless steel socket, bourdon, and tip are used, it is necessary to weld the joints in order to achieve the best corrosion resistance. If the joint were silver soldered, then the corrosion resistance of the pressure element would be only as good as the silver solder. Further, because of the dissimilar metals it is possible that

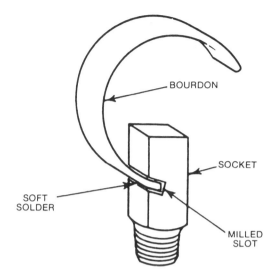

Fig. 3.17 Method of attaching bourdon — slotted.

the joint would be subject to galvanic corrosion. Welding is usually accomplished using the TIG welding process and requires a skilled welder to avoid burning through the thin-walled bourdon tube while attempting to raise the temperature of the relatively massive socket to its melting point. Frequently "welding lips" are provided for the thinner wall bourdons. These lips are formed by machining away a portion of the socket around the spline milled cavity so as to provide a lip that can be more readily melted.

Where very thin-walled type 316 stainless steel bourdons are required, it is sometimes impossible to weld them without burning through. An alternative is to use a brazing alloy high in nickel content. This alloy has better corrosion resistance than the silver solder, but is still not as good as the parent metal, and the application must be carefully considered.

3.12 GAUGE MOVEMENTS

As is true of almost every other component in a pressure gauge, a variety of movement types is available, and individual parts of the same movement type have variations. We will first make some general comments on movements, then identify individual parts of a basic movement and discuss the variations of these parts. Finally we will discuss in brief other types of commonly used movements.

3.12.1 General Comments

The movement of a dial indicating pressure gauge greatly magnifies the small linear motion of the bourdon (measuring element) and converts it into rotary motion so as to produce full travel of the indicating pointer over the dial (usually 270° of arc). To accomplish this, the movement requires the precision manufacture found in watchmaking; dimensional tolerances on the order of 0.0002 in. are sometimes required.

Because of the severe service requirements to which the movement may be subjected, some have rather heavy proportions that give them the appearance of being crude. Actually, the movement gearing and bearings must have a high degree of precision. In contrast to a watch, a pressure gauge does not have a powerful mainspring driving a gear train in one direction. A gauge movement must provide smooth, rolling action in all bearings and on the faces of the gear teeth so that it will respond to minute movements of the elastic chamber in either direction of travel and be repeatable within very narrow limits.

The movement must be as free of friction as possible. This can be accomplished only through precision-generated gear teeth on both the sector and pinion. In addition, the hairspring tension must be correctly set so as to remove the effect of the clearance between the various elements of the amplification train without creating too much friction. Movements generally are not lubricated for several reasons. Gauge cases are usually not sealed so that the internals are subject to atmospheric dust and other debris that would cling to the oiled parts and cause erratic pointer motion. This is especially true of lubrication on the gear teeth. If the lubrication is viscous, the hairspring may not be able to overcome the retarding effect of the lubricant. If the lubrication is light, it will migrate away from the bearings and be useless in a short time. Normal lubricants change their viscosity with ambient temperature; therefore a lubricated movement that operates satisfactorily at room temperature could become erratic at low temperatures. Where exceptionally severe wear conditions exist, some form of dry lubricant, such as molybdenum disulphide is sometimes used. Pointer oscillation caused by external vibration can be alleviated to some extent by the use of a very viscous (30,000 centistoke) silicone compound on the teeth and/or bearings. This is not basically a lubricant but rather acts as a damping fluid. The application of this compound

Gauge Movements

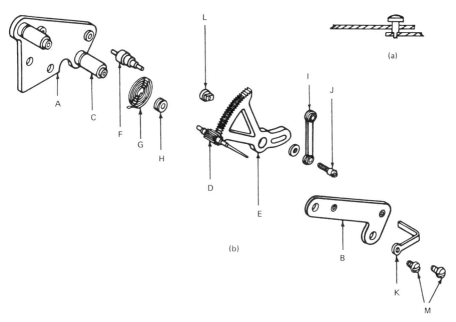

Fig. 3.18 Typical movement components. (a) Detail of link rivet. (b) A, bottom plate; B, top plate; C, column; D, pinion; E, sector; F, arbor; G, hairspring; H, collet; I, link; J, link screw or rivet; K, minimum stop; L, link nut; M, column screws.

will result in slower response time and some inaccuracy due to the inability of the hairspring to completely overcome the drag of the viscous fluid.

3.12.2 Basic Movement

The basic bourdon pressure gauge movement illustrated in Fig. 3.18 is the most common type and is made up of the following components.

1. Top (or front) plate and bottom (or back) plate -- the plates that form the main frame members or "cage" of the movement in which the shaft bearings are located. The top plate is nearest to the gauge dial.
2. Columns (sometimes referred to as "separator posts") — the two members that separate the top and bottom plates. They may be joined to the plates with rivets or screws.

3. Pinion — the small gear that includes the indicating pointer shaft. Its bearings are in the top and bottom plates.
4. Sector (sometimes referred to as "segment") — the part with gear teeth at one end to mesh precisely with the pinion, a hole for the arbor at the center of the gear arc to provide a point about which it rotates, and a "tail" connected through a link to the free end of the bourdon (or other measuring element).
5. Arbor (sometimes referred to as "sector shaft") — the part securely attached to the center hole of the sector, forming a pivot. Its bearings are in the top and bottom plates.
6. Hairspring — this part, shaped like a watch hairspring, takes up the small amount of backlash (that is, the clearance or "play") that exists in the pinion and sector bearings, the gear mesh, and the link and the link rivets. Do not confuse this with the bourdon, which is sometimes referred to as the spring.
7. Collet — a small washerlike part that fits on the pinion shaft and holds the center of the hairspring.
8. Link (sometimes referred to as "segment link") — the member used to transfer motion of the bourdon tip (or other measuring element) to the sector tail.
9. Link screw or rivet — the fastening hardware used to attach the link to the segment and bourdon tip. Grade B gauges normally use a rivet, since the gauge is considered unrepairable. More expensive gauges use a screw and link nut.
10. Minimum stop — an arm that stops the segment at its zero position, in lieu of a zero stop pin on the dial face that stops the pointer.

3.12.3 Variations in Basic Movement

1. Bushed movement. As noted above, there are variations within the movement types. For example, the basic movement may be a "bushed movement," which refers to one having the six bearings points (two pinion bearings, two arbor bearings, and two link bearings) contained in bushings pressed into the movement plates and link. The purpose is to provide increased wear resistance by making the bushing longer than the plate thickness, or by making the bushing from a more wear-resistant material than the plates. By properly selecting the bushing material, it is possible to not only increase the wear life of the bearings but also reduce the friction. It is well known that friction is reduced and wear life is increased if dissimilar materials are used for the mating parts. Commonly used combinations are phosphor bronze and nickel silver, phosphor bronze and stainless steel, and nickel silver and stainless steel. Certain plastics are also used as bearing materials and are often alloyed with Teflon

Gauge Movements 81

(du Pont) to further improve wear life and reduce friction. The combination of plastic bushings and stainless steel rotating components offers an extremely high wear life and low friction. As might be expected, the cost of bushed movements is higher than that of unbushed movements. Therefore the use of bushed movements is generally limited to higher quality process gauges and test gauges.

2. Heavy duty movement. A heavy duty movement is especially constructed for severe pulsation and/or vibration service and may contain bushings (as described above) or it may use extra heavy top and bottom plates to gain the increased bearing length. The segment is likewise usually made from heavier and/or more wear-resistant material. The added weight of the segment causes increased friction and lower sensitivity and therefore is used only when required by the environment in which the gauge operates. A heavy-duty movement may also include a hairspring guard. This is a thin disc attached to the pinion shaft between the segment and the hairspring to prevent the hairspring from becoming entangled in the gear mesh.

3. Corrosion-resistant movement. Corrosion-resistant movements, as the name implies, are made to withstand some particular corrosive environment. Most movements are made of brass, which is satisfactory for atmospheric environments. Where brass is not suitable, the movement is usually made of one of the 300-series stainless steels. The combination of a stainless steel movement with plastic bushings molded from one of the highly corrosion-resistant plastics is frequently found in gauges used in chemical plants and oil refineries.

4. Rotary movement. The term rotary movement refers to a movement mounted in such a way that it can be rotated around the pinion. By rotating the movement, the basic geometry of the gauge layout may be altered to obtain a more accurate translation of the bourdon motion into rotary motion (refer to Secs. 3.4.3 and 7.7.4 for additional information).

If a bourdon is overpressured to the extent that the elastic limit of the material is exceeded, the tip of the bourdon will be permanently moved from its original location, and the pointer will no longer indicate the correct pressure. Rotary movements can be used to re-establish the correct geometry and permit recalibration of the gauge (providing the permanent set does not exceed approximately 10% of the tip travel) and are therefore often used on more expensive gauges that are considered repairable. Rotary movements are usually required on gauges in ANSI class 2A and higher in order to obtain the necessary accuracy. Two variations of rotary movements are shown in Figs. 3.19 and 3.20. The rotary movement shown

82   Chap. 3 Gauge Components

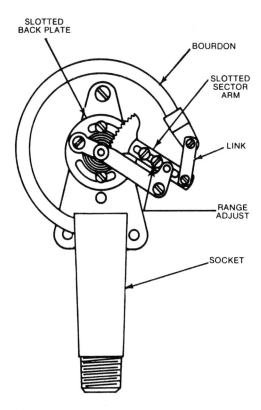

Fig. 3.19 Rotary movement.

in Fig. 3.20 is designed so that it can be disassembled from the gauge for inspection and cleaning, and then reassembled in exactly the same position. In this manner the need to make range and scale shape adjustments following reassembly is eliminated, and the accuracy of the gauge will be unaffected by the movement removal.

3.12.4 Bendable Tail and Slotted Tail Movements

It is not possible to manufacture the bourdon so as to hold the tip travel to an exact amount, and since a given movement has a fixed gear ratio, variations in the tip travel will result in more or less

Gauge Movements 83

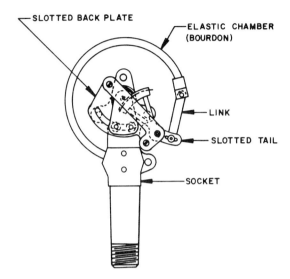

Fig. 3.20 Removable rotary movement.

rotation of the pointer. This variation could be compensated for by varying the length of the dial calibration arc to suit the tip travel of a specific bourdon. However, as we will see later, the dials are mass-produced and it is not practical to match the dial to the bourdon. Therefore, a means to adjust for the tip travel variation must be provided.

The most convenient way of providing the range adjustment is to vary the distance from the arbor to the point of attachment of the link. To explain this in greater detail, assume that the movement has a 10 to 1 gear ratio. If the dial is laid out so that zero to full scale spans an arc of 270° then the sector must move through an angle of 27°. The tip travel required to move the sector 27° depends on the distance from the arbor, as illustrated in Fig. 3.21. It will be noted that either motion a or b will move the sector through the required 27°, and that the closer the point of attachment of the link is to the arbor, the less motion (tip travel) is required. There are two common means to adjust the point of attachment of the link: provide the sector with either a bendable tail (Fig. 3.22) or a slotted tail (Fig. 3.23).

In the bendable tail configuration, the link is permanently riveted to the tail through the hole at the end of the tail. This is a

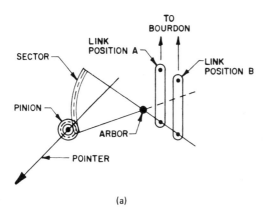

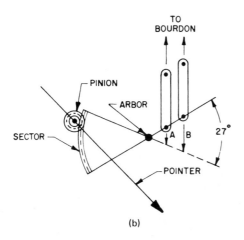

Fig. 3.21 Method of range adjustment. (a) Zero pressure, (b) full-scale pressure.

more economical construction and is found on many lower-priced, general purpose gauges. It has the advantage that the link cannot work loose when the gauge is subjected to rapid pressure pulsations or severe mechanical vibration. It also permits the use of a lighter sector with smaller link bearings, which tends to improve the pointer action. When calibrating, as covered in Chap. 7,

Gauge Movements

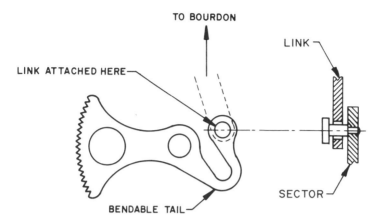

Fig. 3.22 Sector with bendable tail.

adjusting the tail by bending to obtain the required degree of accuracy is not as difficult as it may seem. This construction is generally used on gauges with ANSI grade A or lower accuracy.

In the slotted tail configuration, the link is attached to the sector using a link screw and a special T nut that fits into the sector tail slot so as to prevent turning when the link screw is tightened. The link screw has a shoulder slightly longer than the thickness of the link so that when it is tightened it will permit the link to rotate freely (see Fig. 3.23). This design is used in larger gauges, where the accuracy is ANSI grade 2A or higher, since it provides for a finer adjustment than the bendable tail. It permits the gauge to be disassembled to clean the bearings (particularly desirable when the gauge is used in corrosive atmospheres). Also, the gauge can be calibrated many times without the danger of work hardening and breaking the bendable tail.

A variation of the slotted tail arrangement is shown in Fig. 3.19. In this construction the sector tail is a separate piece attached to the sector with two screws. The link is attached to the tail at a fixed location. To obtain range adjustment, the screws holding the separate piece to the sector are loosened and the separate piece adjusted so that the point of link attachment moves toward or away from the arbor.

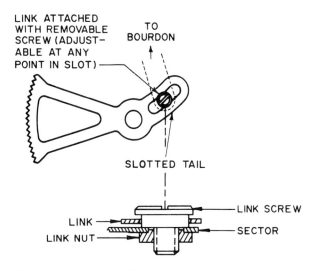

Fig. 3.23 Sector with slotted tail.

3.12.5 Vacuum Movements

One other important variation in basic movements is that used for conventional vacuum gauges. As noted elsewhere, if the internal pressure of the bourdon is below atmospheric pressure the coiling radius of the bourdon will decrease and the direction of the tip travel will be opposite to that generated by pressure above atmospheric pressure. In order to retain a clockwise pointer motion for measuring vacuum, the entire gauge layout must be reversed, including the movement. Therefore, a series of gauges that includes vacuum ranges must include vacuum movements. An advantage of gauges calibrated in terms of negative pressure, with the pointer moving counterclockwise for increasing negative pressure, is that standard movements can be used.

3.12.6 Variations in Basic Movement Components

1. Columns or separator posts. The columns are machined from rod stock or tubing stock. A shoulder is machined on both ends and fits precisely into corresponding holes in the top and bottom plates so as to accurately align the mating bearing holes. Low-cost gauges that are considered nonrepairable have the columns staked

to the plates, permanently trapping the pinion and arbor assembly. Higher-quality gauges use columns threaded on one end so as to receive a machine screw. This permits disassembly of the movement so that it may be cleaned or a damaged hairspring replaced. One of the columns provides a groove into which the outer end of the hairspring is staked.

2. Pinion. Because it is the most critical part of the movement, the pinion is made to a very high degree of accuracy. Careful use of the special machinery for making this part permits minimum bearing clearance. This, in turn, permits closely fitting gear teeth and a closely controlled center to center distance between the two gears to provide longer, more accurate service. Cutting of the gear teeth is a very precise operation. The exact tooth form may differ among gauge manufacturers and therefore interchanging of movement parts from different manufacturers is not recommended. The teeth are coarser in larger gauges, where the movement components are larger, thus enabling the movement to withstand greater wear. As the gauge size is reduced, the movement must be made smaller, with a comparable reduction in the size of the pinion. Because it is not practical to have fewer than 12 teeth on the pinion, a higher diametral pitch must be selected to produce a smaller tooth. Regardless of the size of the gauge, smaller teeth will always produce smoother pointer action with greater sensitivity, because there will be more teeth on a given size pinion. However, smaller teeth are more readily damaged by wear.

3. Hairspring and collet. Some clearance must be left between the sector and pinion bearings, in the gear tooth mesh, and at the link bearings, to insure freedom of operation. The total clearance is quite small, but it will cause a "deadspot" of about 2% of full-scale indication on small gauges. Deadspot refers to a condition wherein there is no pointer motion for a small incremental pressure change or for a reversal of the applied pressure. The hairspring forces all of the moving parts in one direction, effectively removing the various clearances.

The inner end of the hairspring is attached to the pinion by means of a collet and the outer end is attached to one of the columns. The spring is wound with sufficient tension to remove the backlash throughout the entire rotation of the pinion. It is usually mounted so that it will decrease its tension as the pointer moves clockwise with increasing pressure, that is, upscale, but it may be wound in the opposite direction.

The tension of the hairspring must be in keeping with the size and sensitivity of the movement. Too much tension will cause excessive friction and increase the wear on the sector and pinion

Fig. 3.24 Solid link.

teeth. Too little tension will not remove all of the backlash and result in the gauge having a deadspot. Hairsprings are made from a variety of materials such as phosphor bronze, stainless steel, beryllium copper, and Ni-Span C. It is obvious from the above that a properly selected and adjusted hairspring plays an important part in determining the accuracy of a pressure gauge.

4. Link and link screws. The link usually connects to the sector tail and the elastic chamber in the same manner (either with screws or rivets). There are a number of variations in link design, depending on the intended use of the gauge:

(a) Solid link
(b) Slotted link
(c) Bushed link
(d) Slip link
(e) Bimetallic link
(f) Multimetallic link

> (a) Solid link. Used with the bendable tail or slotted tail sector, this type is simply a strip of metal with two bearing holes at a fixed distance apart to accommodate the link screws or rivets connecting it to the tail and the bourdon (or other element) (see Fig. 3.24).
> (b) Slotted link. This is made of two strips of metal, one threaded to receive and secure two locking screws that pass through a slot in the second strip. This permits lengthening or shortening of the link to compensate for changes in the tip position of the elastic chamber and scale shape errors (see Fig. 3.25).

Gauge Movements

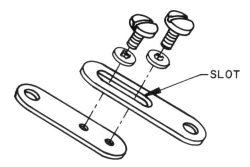

Fig. 3.25 Slotted link.

(c) Bushed link. Bushings may be added to the solid or the slotted link, depending on the type of service. The bushings are fabricated in the same manner as those for the movement plates and generally from the same material.

(d) Slip link. This type (Fig. 3.26) is used wherever it is desirable to have the elastic chamber move a certain distance before it engages the sector tail and causes pointer movement. For example, suppose the first 100 psi on a 500-psi gauge is to be suppressed, that is, cause no pointer movement. The gauge could be equipped with a slip link constructed as shown. The link is made in two parts, one being fastened to the tip and the other to the sector tail. The parts are joined so that they will slide on each other. The correct spacing between the two holes is maintained by a coil spring, which pulls the two parts together against suitable stops. The distance between holes is thus maintained at a maximum but can be shortened by sliding the two parts against the tension of the coil spring. A minimum stop must be employed in conjunction with the link and is set to stop the sector travel in a decreasing pressure direction at 100 psi. Continued travel of the pressure element from 100 psi to 0 psi is taken up by shortening the link while the sector tail remains stationary. On increasing pressure, the link will extend until the stops are encountered, at which time the sector will begin to move. This link is also used to protect the movements of hydraulic gauges used on tensile testing machines

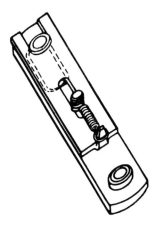

Fig. 3.26 Slip link.

from the shocks resulting from the sudden release of pressure that occurs when the specimen breaks. Such shocks are due to a loss of pressure at a rate more rapid than the movement can normally follow, and without a slip link the bourdon would impose an excessive shock load on the movement.

(e) Bimetallic link. Where it is desirable to have a gauge that will maintain its accuracy over a wide range of ambient temperatures, it is customary to utilize a bimetallic link to correct for zero shift, that is, an error that is constant over the entire scale (see Sec. 6.3.5 for further discussion). These links are made from material having two layers of different metals bonded together (see Fig. 3.27). The metals are chosen so that with an increase in temperature one will expand at a higher rate than the other. If a flat strip of bimetallic material is formed into a U shape, as shown in Fig. 3.27, then a change in temperature will cause the U to open or close depending on whether the high-expanding metal is on the inside or the outside of the U. By properly selecting the materials when the gauge is designed, the curvature can be made to impart a correction to the indication that is just sufficient to compensate for zero shift due to temperature. To obtain complete temperature compensation it is also necessary to correct

Gauge Movements 91

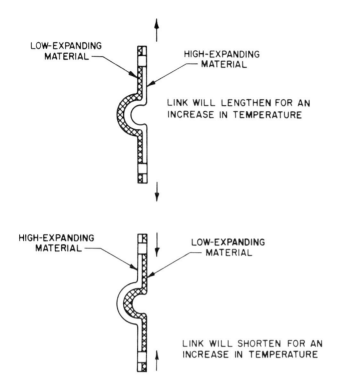

Fig. 3.27 Bimetallic link.

for the change in the spring rate of the elastic chamber (see Secs. 6.3.2 and 6.3.4 for further discussion). Bimetallic links must be specially designed for each type and range of pressure gauge. Therefore it is not possible to use a bimetallic link from one gauge in another one of a different range or design. Bimetallic links should not be used where an overpressure stop is employed since their construction will not permit them to absorb the pull of the pressure-sensitive element when the motion of the sector is blocked.

(f) Multimetallic link. Another type of link used to correct for zero shift due to temperature change is composed of metals having different rates of thermal expansion, as illustrated in Fig. 3.28. With this design, the amount of

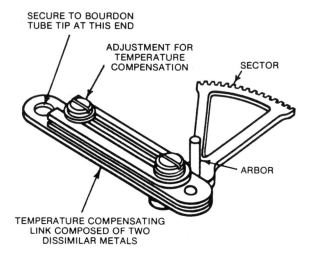

Fig. 3.28 Multimetallic link.

correction can be varied by adjusting the clamp screw. This permits precise setting of the amount of correction for any one gauge and is more universal than the bimetallic link. As with bimetallic links, another means of range correction must be provided. This type of link does not have the limitation with respect to overpressure stops discussed under Sec. 3.12.6.4(e).

5. Minimum and maximum stops. As shown by item K of Fig. 3.18, it is a simple matter to place an L-shaped part under a column screw so that it will stop the sector tail at a certain point of its travel in the downscale direction (clockwise movement of the sector tail). When set in this manner, it is said to be a minimum stop. It has the advantage of preventing the bending of the pointer that can occur if the pressure is suddenly released and the pointer strikes a dial stop pin. A similar stop can be arranged to limit the upscale travel of the sector tail for overpressure protection, in which case it is called a maximum stop. Stops should not impede indicating over the full range of the gauge.

3.12.7 Other Movement Types

1. Helical movement. The movement shown in Fig. 3.29 uses a helically grooved shaft into which the edge of a gearless sector

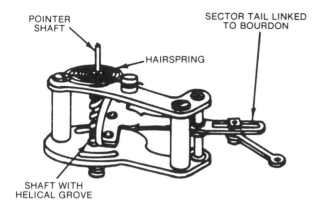

Fig. 3.29 Helical movement.

operates. As the sector moves through its arc, in the same manner as the geared pinion and sector design, it works in the helical groove to rotate the pointer. The sector edge contacting the groove is faced with a graphited, phenolic resin, and the grooved shaft is of polished stainless steel to promote smooth action. Continuous line contact between the sector end and the helically grooved shaft is maintained by the tension of the hairspring.

2. Diaphragm gauge movement. The diaphragm gauge movement shown in Fig. 3.10 is a commonly used movement type. Its operation is discussed in Sec. 3.5. Range adjustments are made by bending one of the pins attached to the rocker arm.

3. Magnetic movement. A special type of movement used in a diaphragm-actuated gauge for low pressures is shown in Fig. 3.30. Linked to the diaphragm center is a calibrated, flat cantilever spring. Mounted on the cantilever opposite the fixed end is a U-shaped magnet, the two poles of which straddle a helically wound flat strip. The magnetic field will maintain a minimum air gap between the helix and the U-shaped magnet. Thus, as the cantilever and magnet are deflected by the motion of the diaphragm, the helix rotates and moves the attached pointer over the scale. The helix may be mounted on jeweled bearings. The device can have full-scale range of 0 to 1.0 in. of water and is commonly used for checking furnace draft and air flow in ducts and in other low-pressure applications.

4. Duplex or dual movement. In Sec. 2.3.4 reference was made to a duplex pressure gauge having the capability of indicating two different pressures. The majority of such gauges has two concentric pointers operating over the same dial arc and requires a special

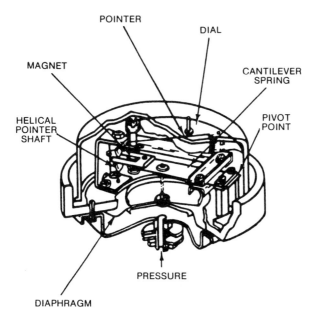

Fig. 3.30 Magnetic movement.

duplex or dual movement. The movement, shown in Fig. 3.31, provides a first pinion that is hollow and a second pinion that projects through the hollow center of the first pinion — similar to the hour and minute hands of a clock. Each pinion is provided with its own bearings and sector/arbor assembly. Independent, pressure-sensitive elements actuate each sector so that the two pointer indications are entirely independent of each other. The hollow pinion, which is necessarily the front pinion of the movement, has large bearings and carries the rear pointer, while the inner or male pinion is necessarily the rear pinion and has a small-diameter bearing and a long thin extension to carry the front pointer. These gauges are used primarily to save panel space, since one gauge can take the place of two. They are sometimes used as a convenient way of comparing two related pressures.

5. Plastic movement. There are several plastic materials from which the various movement parts can be molded and assembled into a complete movement. One difficulty encountered is that the range adjustment cannot be obtained by bending the sector tail, since the

Gauge Movements

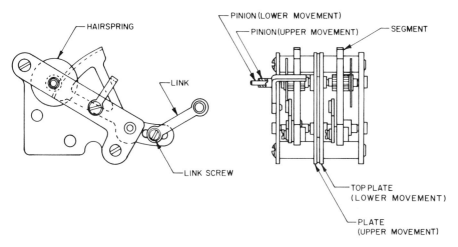

Fig. 3.31 Duplex movement.

plastic is not deformable in the same manner as a metal sector tail. Therefore, it is common to provide either a slotted sector tail or a separate metal tail assembled to a molded plastic sector. Sometimes other metal components, particularly the pinion, are used, and of course the hairspring is also metal, since it would be difficult to fabricate a plastic hairspring.

The principal advantages of the plastic movement are long wear life and, with proper selection of materials, a high degree of resistance to chemical attack. The principal disadvantage is that plastic is not as rigid as metal and is dimensionally more affected by humidity and temperature extremes. It is therefore not used in gauges of class 2A accuracy and higher, although, as previously mentioned, plastic bushings in metal plates are commonly used in the higher accuracy gauges.

6. Gearless movement. As the name implies, gearless movements utilize only linkages to multiply the pressure-sensitive element motion. It is difficult to obtain the large arc of pointer travel provided by gearing, and gearless movements generally provide about 70° of pointer travel. In order to gain as much scale length as possible the center of the pointer rotation is located near the periphery of the gauge, usually the 6 o'clock position, as illustrated in Fig. 3.32. Range adjustment is obtained by bending the link member attached to the pointer shaft; this is similar to the principle used in bendable

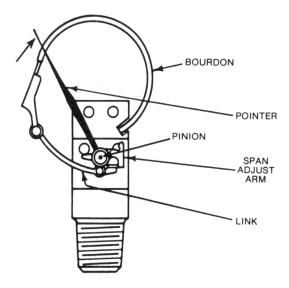

Fig. 3.32 Gearless movement.

tail gear movements. The point of attachment of the link to the bendable member should be as close as possible to the axis of the pointer shaft in order to minimize the amount of motion required of the pressure element. However, practical design considerations require a compromise, and, in general, more motion is required to obtain 70° of pointer travel with a gearless movement than is required to obtain 270° of pointer motion with a geared movement. Therefore, the overpressure and fatigue life of a gearless movement gauge is generally not as good as a geared gauge.

The principal advantage of a gearless movement is that the gearing, which is most vulnerable to wear, is eliminated; therefore gauges with gearless movements are useful in applications where pressure pulsation and vibration are encountered. However, plastic gearing and liquid filling of the gauge to dampen pulsations offer other means to obtain long wear life without sacrificing the greater scale length obtainable with a geared movement.

Gearless movements appear to be less expensive than geared movements, but this is not always the case, since it is usually more difficult to calibrate a gearless gauge.

Dials

7. Miscellaneous. There are other types of movements that may be used for special purposes, having additional links and gears to increase the multiplication, provide nonlinear functions, etc. Other proprietary designs for use with bourdon tube pressure elements are also in use, but a description of these is beyond the scope of this handbook.

3.13 DIALS

3.13.1 General

The dial is the component on which the graduations and numerals (called the scale) are imprinted, together with other specific information for the user such as the units of pressure or intended use. It is usually fastened to the socket in order to provide a unitary structure that will maintain a fixed relationship between the various parts, although it may be fastened to the case or even be imprinted on the underside of the window. Dials are generally made of brass, steel, or aluminum, but where maximum corrosion resistance is desirable may utilize stainless steel. Lighted gauges may use translucent plastic dials.

3.13.2 Graduations

The graduations divide the scale into increments of pressure. In order to improve the readability of the scale, major graduations are used at some fractional interval of the full scale (see Fig. 3.33). The space between the major graduations is divided by minor graduations. Numerals representing the value of the graduation are usually placed at major graduations.

3.13.3 Scales

Dials may consist of a single, black scale on a white background (the most common) or they may be quite elaborate, having multicolor scales, operating zones, customer trademarks, and operating instructions. They may have a basic pressure scale and, in addition, one or more related scales, for example, the equivalent saturation temperature of a specific refrigerant at the corresponding pressure. Some pressure gauges have only one scale, which, while related to pressure, indicates some other variable. For example, a gauge used in conjunction with a hydraulic cylinder may have a scale indicating the force on the piston rather than the actual gauge pressure. (The force is obtained by multiplying the area of the piston

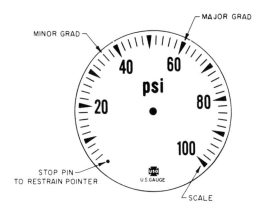

Fig. 3.33 Typical dial.

by the pressure, using compatible units.) Gauges having scales not calibrated in pressure units should contain some note or indication that tells the user the actual pressure range of the gauge in order to avoid possible misinterpretation of the numerical values shown, which in turn might result in misuse of the gauge.

Gauges intended to measure pressure less than atmospheric may have scales calibrated in terms of vacuum or negative pressure (see Fig. 3.34). Dials graduated in terms of negative pressure will have minus signs adjacent to the numerals; increasingly negative numerals proceed in a counterclockwise direction. Dials graduated in terms of vacuum will have numerals increasing in value in either a clockwise or a counterclockwise direction (see Sec. 2.3.2). They will not have minus signs in front of the numerals and will show the word "vacuum" on the dial (for further discussion of negative pressure see Sec. 11.3). The word "pressure" may or may not appear on the dial of negative pressure gauges, it being understood in the same manner as with a gauge measuring positive pressure.

3.13.4 Scale "Take-Up"

1. Explanation of take-up. A very common practice in the layout of scales is the use of a "take-up" (sometimes called "start") at the beginning of the scale. It is important to understand the concept. As an example of a take-up, count the minor graduations in Fig. 3.35, going counterclockwise from the major graduation marked

Dials

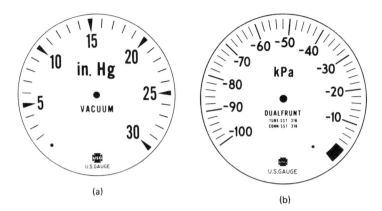

Fig. 3.34 (a) Vacuum dial, and (b) negative pressure dial.

40 to the next major graduation, which has a value of 20 psi, and you will find there are four spaces; that is, each space has a value of 5 psi. Continued counting will show that the first minor graduation has a value of 10 psi, and the pointer is stopped about one space below this graduation. In other words, the value of the stopped pointer position is not 0 psi but is approximately 5 psi, and the pointer is restrained from going to the true zero position. Therefore, at zero applied pressure, the pointer will indicate approximately 5 psi. The difference between true zero and the stopped

Fig. 3.35 Scale with take-up.

pointer position, measured in the units in which the dial is laid out, is called the take-up. In Fig. 3.35 the take-up has a value of 5 psi. Note that the zero numeral and graduation have been omitted. This is in accordance with ANSI B40.1, which states that a zero graduation and/or numeral shall not be permitted at the stopped pointer position on gauges using a zero stop pin or internal stop that prevents free pointer motion to the actual zero pressure position.

2. Reason for take-up. The reason for using a take-up is that the gauge has an accuracy tolerance at zero in the same manner as any other point on the scale. Therefore, if a true zero graduation is used, then at zero applied pressure the pointer could be above or below the graduation by an amount equal to the allowable tolerance. The average user believes that if the pointer is not on zero at zero pressure, the gauge must be faulty, and will either refuse to accept the gauge or attempt to reset the pointer. Resetting the pointer at zero will often destroy the stated accuracy of the gauge. For example, consider a gauge having a range of 0–160 psi as shown in Fig. 3.35 and calibrated to a grade B accuracy (3%-2%-3%). The allowable error over the first and last quarter of the scale is ±4.8 psi, and over the middle half of the scale is ±3.2 psi. Therefore, the allowable error at 0 psi is ±4.8 psi. Suppose the scale had a true zero graduation and the gauge were actually calibrated so that at 0 psi the pointer indicated +4 psi, which is within the allowable tolerance. At 80 psi the pointer could indicate 77 psi, which is also within the allowable tolerance. If the user, noting the +4 psi at zero, moves the pointer to zero, then the indication at 80 psi would no longer be 77 psi but rather 73 psi. The error is now in excess of 4% at midscale, where the gauge is most likely to be used. By selecting a 5 psi take-up, as shown in Fig. 3.35, the pointer will always be against the stop pin regardless of whether the gauge is tested 4.8 psi plus or minus, and it will not be possible to reset the pointer without actually applying a known pressure, which is obviously the only correct way to make any adjustments. Note that the same rationale will apply if, instead of a dial stop pin, an internal stop (i.e., one that restrains some part of the movement) is used to keep the pointer from moving to the true zero position.

3. Reading scales with take-up. It is important to understand that when a take-up is used it is possible that a pressure equal to the sum of the take-up and the allowable tolerance must be applied to the gauge in order to make the pointer move off the stop. Continuing to use the 0–160 psi grade B gauge as an example, suppose the gauge had been calibrated so that at 5 psi the full minus tolerance (3% of span or 4.8 psi) is used. Since the take-up actually restrains the pointer at a true scale value of 5 psi, it would require

Dials   101

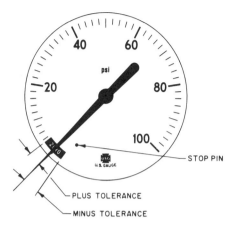

Fig. 3.36 Scale with zero band.

9.8 psi (or about 6.1% of span) to move the pointer off the stop. It is also possible that the pointer will move off the stop pin at a pressure equal to the difference between the take-up and the tolerance. In the above example the pointer could move after 0.2 psi was applied. Therefore a pressure gauge utilizing a take-up at 0 psi should not be used to measure pressure that is less than approximately 10% of the span, because there is a possibility that the pressure to be measured will not be sufficient to move the pointer off the stop. Also, the accuracy of the measurement will be poor. For example, if a 0–160 psi gauge is used to measure 10 psi, the error of the measurement could be as high as ±4.8 psi (3% of 160 psi) or ±48% of the *applied* pressure.

3.13.5 Zero Band

An alternative to using a take-up is the use of a "zero band." The band may take various forms, but essentially it consists of marking off a distance on either side of the true zero pressure equal to the tolerance permitted at zero pressure plus the pointer width. In this manner, when there is a zero pressure applied to the gauge the pointer will be within the zero band. Fig. 3.36 illustrates one form of a dial having a zero band. Note that true zero pressure is centered within the band and that the band extends on either side a distance slightly more than 3 psi or 3% of the 100 psi span. It is usually necessary to provide a stop (either a dial stop pin or an internal stop) to keep the segment from coming out of mesh with the pinion in the event that vacuum is inadvertently applied to the gauge or the gauge is subjected to a severe mechanical shock. If a stop is used it is set so that it will not stop the pointer from moving to a point at or below the end of the zero band. Because the pointer is not stopped it will respond as soon as pressure is applied to the gauge. This is an advantage over the take-up arrangement wherein it is possible that the pointer will not respond until a pressure of as much as the accuracy tolerance plus the value of the take-up is applied to the gauge. A further advantage is that since true zero is at the center of the zero band the accuracy of the gauge at zero is readily apparent by noting the position of the pointer within the band. It should be noted that the pointer could enter the zero band at a pressure equal to twice the numerical accuracy tolerance provided the gauge was tested to the full minus tolerance.

In summary, at zero applied pressure, the pointer will be within the zero band, but if the pointer is within the zero band, the applied pressure is not necessarily zero.

The choice of using a take-up or a zero band is not clear-cut. The take-up is most commonly used, while the zero band is often used on gauges that are specified by instrument engineers. The ANSI B40.1 Standard recommends that no take-up be used on grade 3A and 4A gauges.

3.13.6 Frequency of Graduations

The frequency of the graduations should bear a relationship to the grade of accuracy. For example, it is somewhat misleading to divide the span into 100 increments if the gauge is made to a grade B accuracy. On the other hand, if the gauge is made to a grade 3A accuracy, then 100 increments is not sufficient. Genrally the value of the smallest graduation interval should be one to two times the numerical accuracy of the gauge.

3.13.7 Dial Variations

There are of course many special dials in use, each of which has its own terminology.

1. Mirrored dials have a reflecting band (mirror) around the periphery of the dial in order to reduce parallax errors. Parallax errors are caused by observing the gauge from an angle rather than looking straight on; they result from the fact that the pointer must be some distance above the dial. Using the reflecting band, the observer can position his viewpoint so that the pointer coincides with

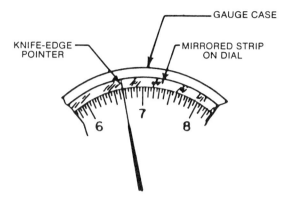

Fig. 3.37 Mirrored dial.

its reflected image, at which time he will be looking straight on. Mirrored dials are generally used in grade 3A and 4A high precision gauges in conjunction with a knife-edge pointer (see Fig. 3.37).

2. Donut dials, as the name implies, are dials printed on an annulus, so that they can be secured on the main dial without removing the pointer. In this way, the gauge may be converted from reading (for example) in psi to kPa or bar.

3. Cupped dials are formed, after being printed, into a dial having a lip extending above the dial face, usually to the underside of the window. The lip may be either vertical or tapered with respect to the dial face. The dial may also be formed into a shallow dish shape (rather than having straight sides), in which case it is called a dished dial. Both cupped and dished dials improve the appearance of the gauge but are more costly.

4. Reverse-printed dials are made with a black (or other color) background with white numerals and graduations.

5. Backlighted dials are usually reverse-printed on transparent or translucent plastic. A lamp is positioned behind the dial so that the numerals and graduations will be lighted and contrast sharply with the dark background. This is one method of lighting automotive gauges.

6. Adjustable dials are attached to the gauge in a manner that permits them to be rotated about the pointer center. They provide a means of making a zero adjustment (see Chap. 7 for explanation of zero adjustment) and are sometimes used as an alternative to adjustable pointers.

7. Dials are often used to dress up the piece of equipment on which the gauge is used, and special finishes, such as brushed aluminum, may be called for. Dials made by etching or stamping the graduations and numerals into brass or zinc sheet have permanent markings and were once popular. However, their cost is very high compared to other methods of obtaining equal permanency, and consequently their use has declined.

3.14 POINTERS

3.14.1 Design Considerations

The pointer of a pressure gauge must be designed to operate in conjunction with the scale of the dial. The length should be selected so that the tip of the pointer does not extend beyond the outer end of any graduation, and it should be not less than the width of a minor graduation from the inner edge of the graduations. It should be bold enough to permit easy reading from a distance. The tail of the pointer is designed to approximately balance the weight of the indicating end of the pointer. This is particularly necessary for grade 2A and higher accuracy gauges in low pressure ranges, since the effect of the unbalance becomes significant. For example, suppose the front end of the pointer is heavier than the tail. As the pressure is increased from zero the effect of the unbalance will be in a counterclockwise direction, increase to a maximum when the pointer is in the 9 o'clock position, decrease to zero when the pointer is in the 12 o'clock position, and increase to a maximum in the clockwise direction when the pointer is in the 3 o'clock position. The unbalance is multiplied back through the gearing and linkage to the bourdon, forcing it out of the position it should be in for the pressure applied, resulting in a corresponding error. Also, if the unbalance is great enough in a direction acting against the hairspring, it may overcome the hairspring and take up the clearance in the gear teeth and bearings in the opposite direction. High-pressure bourdons, which are generally stiffer than low-pressure bourdons, are less affected by pointer unbalance. Likewise, small-diameter gauges are less affected and, in general, permit larger errors; therefore the balance is less critical.

It is best to keep the overall weight of the pointer as low as possible, since heavy pointers load the pinion bearings more and reduce the sensitivity of the gauge, that is, its ability to provide pointer motion for small increments of pressure change.

Pointers are usually made from brass or aluminum and are attached to a bushing for mounting on the pinion shaft (see Fig.

Pointers

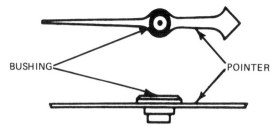

Fig. 3.38 Pointer with bushing.

3.38). The bushing is provided with a central hole having a slight taper (about 2° included angle) that matches the taper on the end of the pinion, and provides a metal-to-metal friction fit with the pinion when it is driven on. Removing the pointer can be accomplished by breaking the friction fit. A pointer lifter, as shown in Fig. 3.39, may be used for easy pointer removal.

3.14.2 Special Pointers

1. Knife-edge pointers. Pointers having a portion of the outer end twisted 90° so that the thickness of the metal becomes the width of the pointer, as shown in Fig. 3.41, are called knife-edge pointers. They are used on test gauges, often in conjunction with a mirrored dial.

2. Adjustable pointers. When calibrating a gauge it is necessary to position the pointer assembly (pointer and pointer bushing) in an exact position on the pinion. For the higher-accuracy gauges very small adjustments are necessary, and it is difficult to position the pointer on the pinion to the required accuracy. To overcome this, adjustable pointers may be provided which permit the pointer to be rotated with respect to the bushing. In this manner the bushing can be driven onto the pinion in an approximate location and the final pointer setting made by means of the adjustment. There are several forms of adjustable pointers, but all operate on the principle of providing relative motion between the pointer and the bushing.

(a) Screwdriver adjustable pointers are made as shown in Fig. 3.40. In this type, the pointer is not rigidly staked to the bushing. A spring washer is interposed between the staking shoulder of the bushing and the pointer. The washer provides sufficient

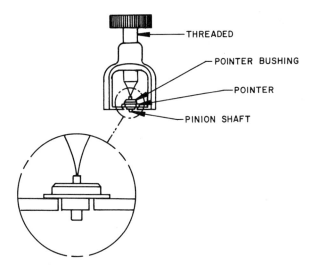

Fig. 3.39 Pointer lifter.

friction to hold the pointer position with respect to the bushing under operating conditions. The head of the bushing is slotted. When it is necessary to make an adjustment, the pointer is held firmly with the fingers, and a screwdriver, inserted in the slot, is turned in the direction of the error. For example, if the pointer indication is plus (error in the clockwise direction), the screwdriver is turned clockwise, thus slipping the pointer on the bushing in a counterclockwise direction.

(b) Geared adjustable pointers are similar to the screwdriver type except that the slot is replaced by fine gear teeth on the head of the bushing as shown in Fig. 3.41. A special pinion

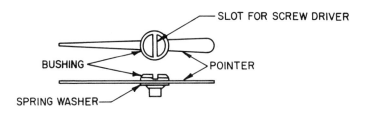

Fig. 3.40 Screwdriver-adjustable pointer.

Pointers

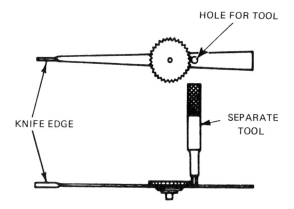

Fig. 3.41 Geared adjustable pointer with knife edge.

tool having mating gear teeth is inserted in a bearing hole in the pointer and meshes with the teeth on the bushing. Turning the pinion tool while holding the pointer provides the adjustment in the same manner as turning the screwdriver, except that the adjusting tool is turned in the opposite direction. Because of the large gear ratio between the pinion tool and the bushings, a more precise adjustment may be made than is possible with the screwdriver slot. Since a special tool is required to make an adjustment, it is possible to control the use of the adjustment feature by limiting the distribution of the tool.

(c) Micro-adjustable pointers are used where an even more precise positioning of the pointer is desired. There are several types in use, two of which are discussed here.

Fig. 3.42 shows one type that uses a worm gear to drive a geared bushing. The bracket holding the worm gear forces the worm gear against the bushing so there is no clearance between the teeth, and the worm therefore acts as a lock on the adjustment, allowing rotation of the pointer with respect to the bushing only when the worm gear is rotated with a screwdriver. Again, the large gear ratio between the worm gear and the bushing permits a very precise adjustment. The amount of adjustment is unlimited.

Fig. 3.43 shows another type of micro-adjustable pointer, using an eccentric carried by the tail of the pointer operating in an arm staked to the bushing. With this type, a locking

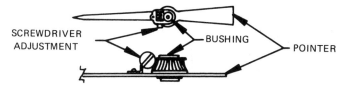

Fig. 3.42 Micro-adjustable pointer — worm gear drive.

screw is provided that must be loosened to make the adjustment and tightened after the adjustment. The amount of the adjustment is limited by the throw of the eccentric. The features of this type are that the parts can be easily fabricated in stainless steel and a positive lock is provided after the adjustment is made. The weight of the adjustment and locking parts help to balance the indicating portion of the pointer; a short tail can thus be used without having to resort to additional balance weights as required by the geared type.

3. Use of adjustable pointers. An adjustable pointer is useful when calibrating a gauge and when making small compensations for liquid head or minor shifts due, for example, to position error. It should not be used to simply set the pointer on the zero graduation (or the middle of a zero band if one is provided) when zero pressure is applied to the gauge. As discussed in Sec. 3.13.4.2, this can result in creating greater errors at another pressure within the gauge span. Further, if the zero shift in calibration is due to a permanent relocation of the bourdon tip, such as may be caused by severe overpressure, then there is usually a change in scale shape in addition to the zero shift. Therefore, a pointer adjustment should be made only after observing scale errors at several points over the gauge span and then deciding the magnitude and direction of the adjustment to provide the optimum accuracy.

3.15 CASE AND RING

3.15.1 Function of Case and Ring

The case, ring, and window assembly enclose the gauge "internals," that is, the measuring element, socket, tip, movement, dial, and pointer. The internals are thus protected to some degree from

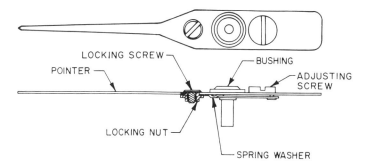

Fig. 3.43 Micro-adjustable pointer — eccentric drive.

mechanical damage, dust, and other debris. The case enclosure may also serve as a means for mounting the gauge to a wall, panel, or piece of equipment.

As discussed earlier in Chap. 2, pressure gauges are supplied in cases representing a wide variety of methods of mounting, materials, types of general construction, etc. Additional information is given in this section, but it would be appropriate for the reader to review Sec. 2.4 so as to avoid repetition here.

3.15.2 Drawn Case — Stem Mounted

The most common type of case is made from a steel stamping painted to provide corrosion resistance (see Fig. 3.44). These are often referred to as drawn cases, stamped cases, or simply steel cases. The same type is also made from brass, aluminum, and stainless steel. Brass cases are usually polished and lacquered or chrome-plated; aluminum cases are anodized or painted. Stainless steel, of course, provides the maximum corrosion resistance but is the most expensive. Polished brass cases are expensive, but provide a very attractive appearance. To reduce the cost, a brass-colored lacquer can be applied to steel or aluminum cases.

Stamped cases are generally provided in sizes from 1 1/2 to 4 1/2 in. and are available on gauges having low, center-back, and low-back connections. The great majority of the grade B general purpose commercial gauges use stamped steel cases because they offer good protection and corrosion resistance and are the least expensive.

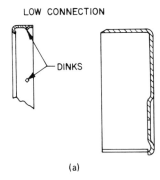

(a)

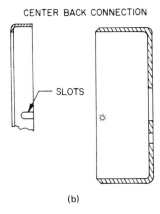

(b)

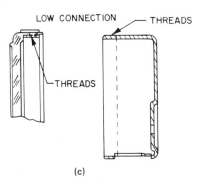

(c)

Fig. 3.44 Drawn stem-mounted cases and rings. (a) Friction ring, (b) slip ring, (c) threaded ring.

Case and Ring

3.15.3 Rings for Drawn, Stem-Mounted Cases

The ring is usually made of the same material as the case although it is somewhat thinner. Drawn cases may use any of the following ring styles.

    1. Friction ring. This ring is stamped and painted in the same manner as the case and is the most commonly used ring. It is pressed onto the case and held by friction. Three or more inwardly projecting dinks are formed in the skirt of the ring to improve the holding force.
    2. Slip ring. This is similar to the friction ring except the dinks are omitted so that the ring slides easily onto the case. The ring is held to the case by providing two slots in the skirt of the ring through which screws are threaded into the case periphery. Slip rings are used when it is desirable to easily remove the window, such as when making use of an adjustable pointer.
    3. Threaded ring. A thread is machined, rolled, or stamped onto the outside diameter of the case. A ring machined from tubing, having a knurled o.d. and a matching thread on the i.d. is then screwed onto the case. See Sec. 3.16.1.4 for combination threaded ring and window molded from plastic.

3.15.4 Drawn Cases — Panel-Mounted

These cases are manufactured in the same way and from the same materials as the stem-mounted drawn cases. However, they are made with a front flange, that is, a lip around the front edge of the case which enables the gauge to be mounted into a panel from the front so that most of the gauge projects behind the panel. Gauges having cases of this type are referred to as flush-mounted or panel-mounted and are made with a center-back or low-back connection. There are two basic means for mounting these cases in the panel. One is to provide the case with a wide flange that contains three holes, by means of which the gauge can be bolted to the panel (see Fig. 3.45). The size of the panel opening and the diameter of the circle on which the three mounting holes are located have been standardized, as shown in Fig. 3.46. The second method of mounting is by means of a U clamp. In this method the case is provided with a narrow front flange, and two threaded studs are attached to the back of the case. After the gauge is inserted in the panel hole, a U-shaped clamping bracket is assembled over the threaded studs and held in place by two nuts (see Fig. 3.47).

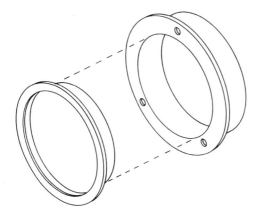

Fig. 3.45 Drawn panel-mounted case and ring.

Gauge size	Bolt circle dia.	Case bolt hole dia.[a]	Panel opening
1 1/2	1.91	0.13	1.65
2	2.56	0.16	2.19
2 1/2	3.13	0.16	2.81
3 1/2	4.25	0.22	3.81
4 1/2	5.38	0.22	4.94
6	7.00	0.28	6.50
8 1/2	9.63	0.28	9.00
12	13.50	0.28	12.62
16	17.00	0.28	16.50
50 mm	60 mm	3.0 mm	Gauge size is outside diameter of case
80 mm	95 mm	4.8 mm	Add. desired clearance
100 mm	116 mm	4.8 mm	

[a]Dimensions in inches unless otherwise noted; three mounting holes located at 12, 4 and 8 o'clock.

Fig. 3.46 Mounting dimensions.

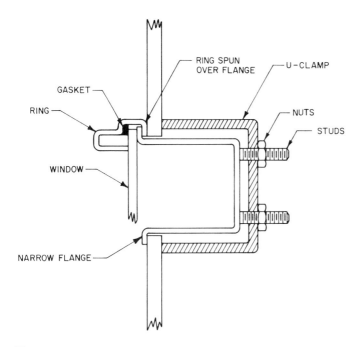

Fig. 3.47 U-clamp panel-mounted case (automotive).

3.15.5 Rings for Drawn, Panel-Mounted Cases

Gauges mounted to the panel by means of a wide front flange utilize rings secured to the i.d. of the case by friction or threading (see Fig. 3.48). Gauges mounted by means of a front flange and U clamp may use rings similar to the wide flange cases or they may have rings spun over the narrow flange as shown in Fig. 3.47. Spun-over rings are commonly used for panel-mounted automotive gauges in conjunction with a gasket that seals the window to the case, thus providing resistance to the entrance of dust and moisture. Spun-over rings of this type must be cut apart to be removed. Drawn cases having a threaded ring as described in Sec. 3.15.3.3 are sometimes provided with studs and a U clamp for panel mounting. In this instance the U clamp pulls the gauge into the panel against the threaded ring. This action tends to push the threaded ring off the case; therefore well-matched threads are required. The ring cannot be removed while the gauge remains in the panel, as is possible with a ring fitted to the i.d. of a flanged case.

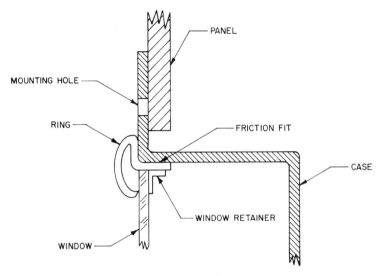

Fig. 3.48 Ring for flange-mounted case.

3.15.6 Cast Cases

Process gauges, industrial gauges, and test gauges are usually provided with a more rugged case made of die-cast aluminum or a corrosion resistant plastic. They are almost always 4 1/2 in. in diameter, although there is some usage in 3 1/2- and 6-in. sizes. The die-cast aluminum cases are supplied in two basic forms: back flange for wall or stem mounting (Fig. 3.49) and front flange for panel mounting (Fig. 3.50). The bolt circle diameter and the diameter of the bolt holes in the back flange cases are the same as the front flange cases (Fig. 3.46). Panel mounting is accomplished by bolting the front flange of the case to the panel. U clamps are not used, because the case is too large and heavy. Die-cast aluminum cases with a baked-paint finish offer good corrosion protection and the ability to withstand considerable abuse. The stem-mounted version is usually stocked with a low connection, although a low-back connection is available. The panel-mounted version is always low-back connected.

Chemical and petroleum processing plants frequently specify gauges having a "turret case." The name derives from its shape (see Fig. 3.51). It is made from a filled phenolic resin or other

Fig. 3.49 Cast case – back flange.

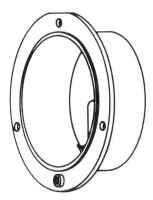

Fig. 3.50 Cast case – front flange.

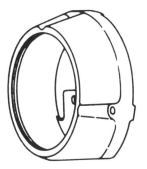

Fig. 3.51 Turret case.

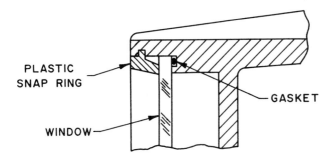

Fig. 3.52 Turret case window seal.

corrosion-resistant plastics. Phenolic is a thermoset resin that has excellent corrosion resistance and a temperature rating of 300°F. These gauges are intended for stem mounting. An adapter ring is available to enable the gauge to be panel-mounted but is infrequently used.

3.15.7 Rings for Cast Case Gauges

Die-cast aluminum cases for stem mounting generally use die-cast aluminum rings. They may be secured to the case by threading or a bayonet lock arrangement. The window is gasketed and secured to the ring so as to provide a seal when assembled to the case. The cases may also use a plastic ring that snaps into an external groove on the case or a metal slip ring. These rings are less expensive than the cast rings but do not provide the same degree of window seal as the threaded rings. Turret cases are usually designed with a grooved window support and require only a simple snap ring to hold the window to the case (see Fig. 3.52). Cast case gauges for panel mounting are often furnished with a hinged ring held to the case by a single thumbscrew. If it is necessary to make a pointer adjustment, the ring can be easily removed while permitting the gauge to remain in the panel (see Fig. 3.53).

3.15.8 Drawn Stainless Steel Cases

Some process applications require that the case and ring be made of type 300 series stainless steel to provide adequate corrosion

Case and Ring 117

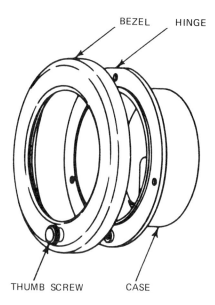

Fig. 3.53 Hinged ring.

resistance. A typical case of this design is shown in Fig. 3.54. A gasketed stainless steel bayonet ring is usually provided (see Fig. 3.55).

3.15.9 Solid Front Cases

The cases discussed in Secs. 3.15.2–3.15.8 are classed as "open front" or "solid back" cases. That is, the periphery of the case and the back of the case are integral portions, and the gauge internals, including the dial, are mounted through the open front to the back of the case. Another series of cases, called "solid front" cases, is similar to the solid back cases except that they are made having a wall integral with the case interposed between the pressure-containing element and the dial (see Fig. 3.56). This requires that the gauge internals, less the dial and pointer, be mounted through the opening in the back of the case and mounted to the integral wall. This opening is then closed with a cover plate, in such a manner that if pressure should build up within the case due to a rupture of the pressure-containing envelope,

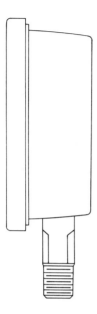

Fig. 3.54 Stainless steel process gauge case.

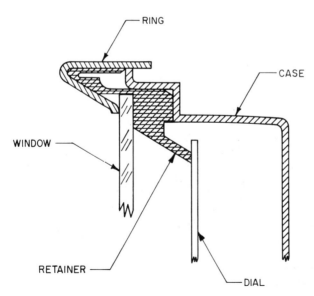

Fig. 3.55 Stainless steel bayonet ring.

Case and Ring 119

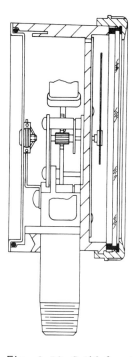

Fig. 3.56 Solid front case.

the cover plate will be pushed out of position, thus relieving the case pressure in a direction away from the observer. The movement pinion shaft projects through the integral wall, and the dial and pointer are mounted on the side of the wall facing the observer. Most process gauges are now supplied in the solid front configuration and in fact are often specified by chemical and petroleum processors. They can be obtained in the same case styles as the open front cases described in Secs. 3.15.6−3.15.8, that is, cast aluminum back flange and front flange, turret, and stainless steel. Solid front construction can also be obtained in drawn stainless steel cases in 63 mm and 100 mm sizes.

3.15.10 Miscellaneous Enclosures

Several special forms of case, ring, and window are supplied for special applications: edge reading gauges, plastic cases, internally lighted gauges, and sealed cases.

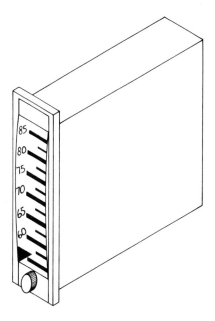

Fig. 3.57 Edge gauge.

1. Edge-reading gauges. Enclosures for edge-reading gauges are generally rectangular in shape (see Fig. 3.57) and may be stampings, die castings, or molded plastic. Since edge gauges are intended to conserve frontal space when mounted in a panel, the case must provide the panel mounting means.

2. Plastic cases. Several varieties of all plastic cases for commercial gauges have been brought into common usage. They are generally more expensive than metal cases but provide increased corrosion and weather resistance compared to painted steel. Two forms are shown in Fig. 3.58. Both forms are available in a low connection as well as the center-back connection shown.

3. Internally lighted gauges. Internally lighted gauges are required by the automotive and off-highway vehicle industries. The source of illumination is an incandescent lamp mounted through a hole in the back of the case. Lighting is accomplished in one of two ways: (1) by use of translucent plastic dials, which allow light to pass through and silhouette the indicia (usually a black background with lighted indicia) or (2) by use of an opaque dial

Case and Ring

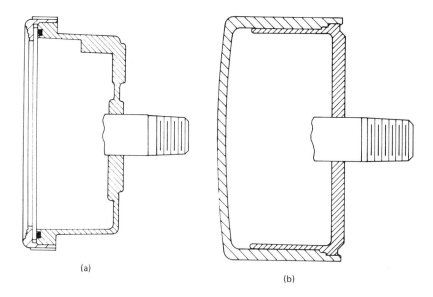

Fig. 3.58 Plastic cases. (a) Threaded ring, (b) snap-on ring and window.

surrounded by a plastic reflecting ring mounted on the periphery of the dial so as to flood light across the dial face. The pointers are most often lighted by flooding, but backlighted pointers are also available.

Where several gauges are used in a cluster or common panel, such as in automobiles or trucks, a common case can be provided into which all of the gauges are mounted. This provides an economical arrangement, since individual cases, rings, and windows are not required and a single lamp can provide illumination for two or more gauges.

4. Sealed cases. There are certain applications where cases sealed against the entrance of water and dust are desirable. Such cases can be provided, but it must be strongly emphasized that some form of case pressure relief is necessary. There is always the possibility that the pressure-containing envelope will develop a leak during use. If the case is completely sealed, then the pressure will build up inside the case. This had two adverse effects. First, the pressure indication will be affected by the amount of the case pressure, wince the bourdon reacts only to the difference

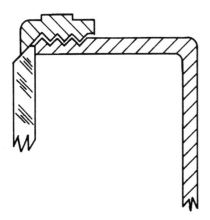

Fig. 3.59 Beveled window and threaded ring.

in the pressure on its internal surface and the pressure on its external surface. Second, and more important, the enclosure (most often the window) can be ruptured, with the possible result of injury to the observer or damage to surrounding property. Various pressure-relief devices are available to meet the application requirements. Users should never seal the enclosure openings or venting devices provided by the manufacturer.

3.16 WINDOWS

3.16.1 General

The transparent cover through which the indicated pressure is observed is called the window. It is often referred to as the glass or crystal, but this is confusing since it is sometimes made of a transparent plastic. Windows are furnished in a variety of configurations: flat, beveled, crowned, and molded plastic.

1. Flat windows. Flat plastic windows may be cut or stamped from plastic sheet, or they may be molded. Cellulose acetate butyrate, polycarbonate, and acrylic are commonly used materials. Flat glass windows are most often cut from plain glass sheet (glazing sheet), but they may be made from laminated safety glass or tempered glass, depending on the requirements. Plastic offers resistance to breakage but is subject to chemical attack and abrasion.

Windows

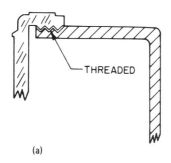

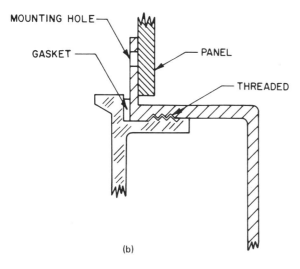

Fig. 3.60 One-piece ring and window. (a) Stem mount, (b) panel mount.

Glass offers high resistance to chemical attack and abrasion but is subject to breakage. Laminated glass offers greatly increased resistance to breakage but is too expensive for general use. The majority of commercial gauges are furnished with flat glass windows cut from plain glass sheets.

2. Beveled windows. Beveled windows are similar to flat windows except that they are made of thicker material and are beveled on the outer edge. They are generally used with threaded rings and cases (see Fig. 3.59). Plastic beveled windows are usually

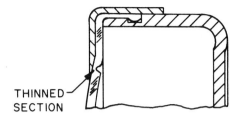

THINNED SECTION

Fig. 3.61 Pressure-relief window.

molded from the same materials as the flat plastic windows and follow the traditional design of beveled glass windows.

3. Crowned windows. Crowned windows are used for aesthetic reasons. Crowned glass is very expensive and subject to breakage. Crowned windows in plastic are readily molded and are no more expensive than flat plastic windows. Because they are crowned, they can be molded somewhat thinner than flat windows, since their shape gives them added strength.

4. Molded plastic windows. Molded plastic windows allow the gauge designer to incorporate features not available with glass. For example, the window can be molded with an integral bezel ring having an internal thread for assembly to a threaded case. This eliminates the need for the separate metal ring used with beveled glass windows. This configuration is called a "one-piece ring and window" and is usually molded from polycarbonate. Another configuration of a one-piece ring and window is used for panel-mounted gauges. This type threads into the i.d. of the case, which permits easy removal of the window without removing the gauge from the panel (see Fig. 3.60).

Windows molded from cellulose acetate butyrate can be provided with a thinned-down section adjacent to the periphery. When used in conjunction with sealed cases, the window will tear at the thinned-down section in the event that a leak in the pressure-containing elements results in a buildup of internal case pressure. This technique is used on fire extinguisher gauges where a watertight case is required (see Fig. 3.61).

4
Liquid-Filled Gauges

4.1 INTRODUCTION

The term "liquid-filled gauges" is used to describe gauges having enclosures that are sealed and filled with a clear viscous fluid. The purpose of filling is threefold. First, rapid motion of the moving components of the gauge, such as may occur when the gauge is subjected to pressure pulsations, mechanical shock, and vibration, will be damped by the surrounding viscous fluid, without impairing the ability of the gauge to respond to more slowly changing pressure inputs. Second, because the gauge internals are surrounded by a compatible fluid, corrosion due to the ambient atmosphere is eliminated. Third, the fluid acts as a lubricant for the moving components and increases wear life. While more expensive than the standard dry gauges, the increased life under adverse conditions can often justify the added cost. Although liquid-filled gauges have been available for more than 50 years, the usage has greatly increased in the past 10 years, so it is appropriate to examine their construction in some detail.

4.2 INTERNAL CASE PRESSURE

There are several characteristics inherent in liquid-filled gauges that should be considered when specifying their use. The gauge enclosure must be sealed in order to retain the liquid fill. In any gauge having a rigid, sealed enclosure, the pressure being measured is referenced to the internal case pressure, not to atmospheric pressure. The indicated reading is influenced by temperature, because both the air and liquid contained within the case expand with increasing temperature and contract with decreasing temperature. Unless some means is provided to allow for these volumetric changes, the internal case pressure will increase with increasing temperature and decrease with decreasing temperature. The effect of this is to change the pressure indication by the same amount, but in the opposite direction, as the case pressure changes. For example, if a non-filled gauge is sealed at standard atmospheric pressure and the internal case temperature increases from 70 to 140°F, then the internal case pressure will increase by about 2 psi. The indication of pressure will decrease by 2 psi because the pressure-sensitive element (bourdon, bellows, or diaphragm) responds to the difference in pressure across it, so a 2 psi increase on the outside of the element has the same effect as a 2 psi decrease on the inside of the element. This represents a 6.7% error if the gauge has a span of 30 psi, but only a 0.2% error for a gauge having a span of 1000 psi. The change in the pressure of a gas within a sealed constant volume enclosure due to ambient temperature changes can be readily calculated using the following relationship, assuming that the case volume remains constant:

$$P_2 = \frac{P_1 T_2}{T_1}$$

where P_1 is the absolute pressure at which the enclosure was sealed, usually 14.7 psia (atmospheric pressure), P_2 the absolute pressure resulting from the temperature change, T_1 the absolute temperature prior to the change in temperature, T_2 the absolute temperature after the change in temperature, and $P_2 - P_1$ the pressure change. Both T_1 and T_2 must be written in "degrees absolute." To obtain degrees absolute, add 460 to the temperature expressed in degrees Fahrenheit or add 273 to the temperature expressed in degrees Celsius (centigrade). In the case of a liquid-filled gauge, the change in pressure will be greater than for a sealed dry gauge because the liquid will expand with increasing temperature and reduce the volume of the air bubble, further increasing the pressure within the case. If the temperature decreases, the liquid will contract and increase the size of the air bubble, further decreasing the pressure within the case. The pressure change due to the volume change of the liquid will depend on the temperature coefficient of the filling fluid, the change in volume of the enclosure, and the ratio of the volume of liquid to the volume of the air bubble. Using the previous example, if the liquid expanded so as to reduce the volume of the bubble by 20%, then the increase in case pressure would be about 6 psi instead of 2 psi, or an error of 20% for a gauge having a full scale range of 30 psi. The calculation is made as follows:

The change in pressure of a gas within a sealed enclosure due to a change in volume of the enclosure is given by the relationship

$$P_2 = \frac{P_1 V_1}{V_2}$$

where P_1 is the absolute pressure at which the enclosure was sealed, usually 14.7 psia (atmospheric pressure), P_2 the absolute pressure resulting from the volume change, V_1 the original volume, V_2 the final volume expressed in the same units as V_1, and $P_2 - P_1$ the pressure change.

Combining the effects of both a temperature change and a volume change on a gas gives the relationship

$$P_2 = P_1 \frac{V_1}{V_2} \frac{T_2}{T_1}$$

where P, V, and T are defined in the same manner as above. Assuming an increase in temperature from 70 to 140°F and a volume reduction in the air bubble of 20% (i.e., final volume is 80% of original volume), the calculation would be

$$P_2 = 14.7 \times \frac{1}{0.8} \times \frac{600}{530} = 20.80 \text{ psia}$$

The total increase in pressure is therefore

20.8 psia - 14.7 psia = 6.1 psi

4.3 RELIEVING INTERNAL CASE PRESSURE

There are two basic ways to avoid the error in indication resulting from case pressure change. One is to provide a case vent to atmosphere that can be closed off during shipping or handling and

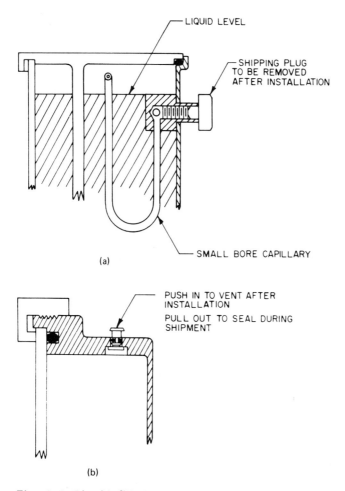

Fig. 4.1 Liquid-filled gauge vented to atmosphere: vents (a) and (b).

readily opened after the gauge is installed and there is no longer a possibility of the fill fluid running out of the vent. The vent must communicate directly with the air bubble when the gauge is in the installed position in order to prevent loss of fluid. Vents of this type are shown in Fig. 4.1. The second basic way is to provide the enclosure with a flexible wall that will change the case volume

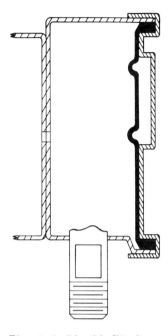

Fig. 4.2 Liquid-filled gauge with pressure-compensating flexible wall.

with the change in case pressure, thus minimizing the pressure change within the enclosure. One method of doing this is shown in Fig. 4.2. In this method, the filling liquid is contained in the enclosure by means of an elastomeric member.

4.4 CHOICE OF FILL FLUID

Any viscous fluid that is noncorrosive to the gauge internals and will not attack the elastomeric seals may be used. The common fill fluids are glycerine and silicone. Commercially pure glycerine becomes too viscous at approximately 40°F and is therefore usually mixed with 10 to 20% water. Silicones are more expensive but are more inert and exhibit small change in viscosity over a wide temperature range. For these reasons silicone is frequently used as the liquid fill. The optimum viscosity may vary depending on the application, but 200 to 300 centistokes is in general use.

4.5 CASE VENTING OF LIQUID-FILLED GAUGES

The atmospheric venting arrangements discussed in Sec. 4.3 may not be large enough to relieve the internal case pressure buildup that will result if the pressure-containing element should suddenly rupture, especially if the gauge is used in a high-pressure gas application. Therefore, some additional means of venting the case is necessary. With solid back, liquid-filled cases, various types of rubber plugs or spring-loaded poppet valves are used. With solid front cases, the entire back or some portion of it can be designed to lift away from the case. The rate of venting required will depend on the type of pressure element failure (e.g., a pinhole leak or an explosive rupture), the pressure medium, and the pressure range. As noted in Sec. 3.15.10.4, the venting arrangement must be maintained in a "free to operate" condition and must not be locked, painted shut, or otherwise interfered with. Additional discussion of case venting is given in Secs. 5.3 and 10.5.9.

4.6 COMPATIBILITY OF FILL FLUID

Consideration must be given to the possibility that the medium whose pressure is being measured will leak into the interior of a filled gauge and mix with the fill fluid. Certain highly oxidizing chemicals, such as nitric acid, chlorine, hydrogen peroxide, and oxygen, may react explosively with the fill fluid. If such chemicals are to be applied to the gauge, glycerine and silicone fill fluids should not be used. Special highly stable fluids such as Fluorolube (Hooker Chemical Co.) are recommended for the fill fluid.

4.7 CASE STYLE

Because of the need for a sealed case, liquid-filled gauges are not available in the wide variety of sizes, case styles, and connection locations offered by the nonfilled gauges. Cases made of stainless steel in 63- and 100-mm sizes are most common. The low-connected gauges are usually stem mounted and the back-connected gauges are provided with a front flange for panel mounting. Both solid front and solid back construction are available. Plastic case gauges in the 2 1/2-in. size with either back-connected U-clamp mounting or low-connected stem mounting are also offered. In the process gauge line the 4 1/2 in. solid front turret case, having a low connection, is the most popular.

4.8 PRESSURE ELEMENTS

All the materials used for pressure elements of dry gauges are available for liquid-filled gauges.

5
Gauge Accessories

5.1 INTRODUCTION

5.1.1 Types of Accessories

This section discusses various devices that may be used in conjunction with pressure gauges to improve their ability to withstand adverse environmental conditions and to broaden their usefulness. The following accessories are generally adaptable to all types of gauges:

1. Diaphragm seals — used to isolate the pressure-measuring element from the pressure medium
2. Case pressure relief devices — used to vent case enclosure
3. Pulsation dampers — used to reduce the magnitude of pressure pulsations
4. Gauge cocks — used to close off pressure source and/or to throttle down pressure pulsations
5. Siphons — used in the pressure line to provide a liquid seal against high-temperature steam or other condensable vapors
6. Bleeders — used to permit flushing, liquid filling, or draining of the pressure-sensitive element
7. Heaters — used to protect lines and gauges from solidification of the pressure medium
8. Maximum and minimum pointers — used to give an indication of the maximum or minimum pressure applied over a given period of time

9. Alarm contacts — used to complete an electrical circuit at a predetermined high or low pressure
10. Automatic shut-off valves — used to protect the gauge against excessive overpressure

5.2 DIAPHRAGM SEALS

5.2.1 Intended Use

Diaphragm seals are intended for use where:

1. The pressure medium would corrode the socket and/or the pressure-sensitive element
2. The pressure medium contains suspended solids or is sufficiently viscous to clog the pressure connection or pressure sensing element
3. The pressure medium might freeze or solidify in the pressure connection or pressure-sensing element due to low ambient temperature or following process shutdown
4. The pressurized system requires flushing, such as may be required to change the process medium
5. The pressurized system must be "sanitary," that is, contain no crevices or blind ports that cannot be readily cleaned and sterilized by flushing, usually with steam

5.2.2 Construction

Diaphragm seals utilize a thin, highly compliant diaphragm between the process medium and pressure-containing internals of the gauge. The volume on the gauge side of the diaphragm is completely filled with a compatible fluid. The pressure to be measured is applied to the opposite side of the diaphragm and is transmitted to the pressure-sensitive element through the fill fluid. Fig. 5.1 represents a typical diaphragm seal having a lower housing which can be threaded to the process piping. The gauge may be attached directly to the diaphragm seal or through a suitable length of capillary tubing if remote mounting of the gauge is desirable.

5.2.3 Importance of Adequate Filling

It is important that the volume on the gauge side of the diaphragm be completely filled with fluid. Any air trapped in this volume will be compressed as pressure is applied to the seal, requiring additional deflection of the separating diaphragm. While the diaphragm

Diaphragm Seals

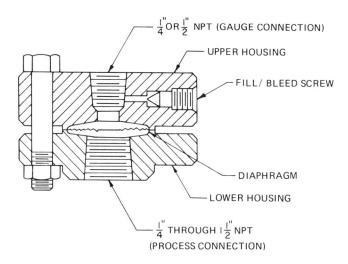

Fig. 5.1 Diaphragm seal – thread attached.

is made to be as compliant as possible, only a limited amount of motion can be obtained without requiring substantial pressure to deflect the diaphragm. Whatever pressure is required to deflect the diaphragm is not transmitted to the fill fluid, so the pressure indicated on the gauge will be less than the applied pressure. If the seal is used to transmit negative pressure (vacuum), air in the filled system becomes even more critical, since it must be expanded, rather than compressed, and a substantial motion of the diaphragm is required to create a high vacuum in the filled system. For this reason it is customary to limit measurement of pressure below atmospheric to 5 psia for metallic diaphragms. Elastomeric diaphragms, because of their higher compliance, will permit indicating to 2.5 psia using materials such as Teflon (du Pont) and 0.5 psia using more flexible materials such as Viton (du Pont). These limitations are based on the very small changes in the volume of the bourdon tube over its pressure range. If the pressure-sensitive element is a bellows or diaphragm stack, requiring a large volume change, there will be other limitations. Another important reason the diaphragm requires a high degree of compliance is that it must accommodate the change in volume of the fill fluid with temperature. This requirement is similar to that discussed in Sec. 4.2. As is true in liquid-filled gauges, temperature errors due to expansion or contraction of the fill fluid are more critical in low-pressure gauge applications than in high-pressure gauge applications.

5.2.4 Materials for Diaphragm Seals

Since the upper housing is wetted by only a noncorrosive fill fluid, it is generally made of plain steel, suitably plated to prevent external rusting. For more critical applications, type 316 stainless steel is available. As might be expected, the diaphragm is available in a wide range of corrosion-resistant materials, since it is wetted by the process fluid. It must be highly resistant to attack, because a corrosion rate of a few thousandths of an inch per year would be very detrimental to such a thin, flexible member. Metallic materials available include various types of stainless steel (especially 316L), Monel, Inconel (Huntington Alloys, Inc.), nickel, Hastelloy (Cabot Corp.), tantalum, and titanium. Elastomeric diaphragm materials include Teflon, Viton, and various synthetic rubbers. The lower housings are likewise available in many corrosion-resistant materials. Type 316 and type 304 stainless steel are frequently used, as well as Hastelloy, Monel, and Inconel. For those applications where the problem is not corrosion but a process fluid that might solidify or is too viscous, a lower housing made of plain steel may be the least expensive selection. Permeation through elastomeric diaphragms may be a problem in some applications, and their use should be discussed with the manufacturer. For some applications, both upper and lower housings may be fabricated from plastic such as PVC, polypropylene, or nylon.

5.2.5 Fill Fluids

Fill fluids should be selected so that in the event the diaphragm fails or a leak develops in the seal, mixing the fill fluid with the process fluid will not result in an explosion or other catastrophic failure, including contamination of the process medium. Contamination may be particularly serious in food processing applications. Other characteristics of the fill fluid that must be considered are its compatibility over the operating temperature range, including stability, vapor pressure, viscosity, and rate of expansion. Some of the more commonly used fill fluids and their properties are as follows:

1. Silicone DC-704

    (a) Usable from +70°F to +600°F. Not recommended for use below +70°F because crystallization can occur.

(b) Will react violently with strong oxidizing agents such as (but not limited to) nitric acid, oxygen, chlorine, and hydrogen peroxide. Therefore, it should not be used for applications where such agents are present, as there is always the possibility that the seal will leak or be ruptured, allowing the fill fluid to mix with the process fluid.
(c) Low change in viscosity with temperature.

2. Silicone DC-200

    (a) Usable from -60°F to +300°F.
    (b) Should not be used for applications involving chlorine. Use with other oxidizing agents, such as oxygen, nitric acid, and sulfuric acid, should be discussed with the seal manufacturer, since potential reaction is a function of pressure, temperature, and the concentration of the pressure medium.
    (c) Less expensive than DC-704.
    (d) Low change in viscosity with temperature.

3. Fluorolube FS-5

    (a) Usable from -50°F to +500°F.
    (b) Chemically stable with respect to oxidizing agents.
    (c) Reacts violently with diamines and aluminum compounds.
    (d) Expensive compared to other fluids.

4. Instrument oil

    (a) Usable from +30°F to +300°F.
    (b) General purpose filling fluid.
    (c) Larger change in viscosity than with the silicones. May cause sluggish indication in cold.
    (d) Least expensive of the fill fluids.

5. Sanitary filling fluids for food applications. Liquids such as water, vegetable oil, and pharmaceutical grade mineral oil may be used.

    (a) Usable temperature range is limited. Check with manufacturer.
    (b) Not for use with strong oxidizing agents (see discussion under DC-704 above).

138   Chap. 5 Gauge Accessories

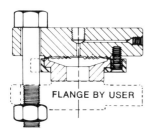

Fig. 5.2 Diaphragm seal — flange attached: type RC. (Courtesy of Mansfield & Green Division, AMETEK, Inc.)

5.2.6 Construction Variations

While the principle of separating the process fluid from the pressure-sensitive internals by means of a compliant diaphragm remains the same, there are many variations in seal construction, each having a specific application or function.

    1. Mounting arrangements. In addition to mounting by threading to the process piping (as illustrated in Fig. 5.1), seals are also supplied for other mounting arrangements as shown in Figs. 5.2 – 5.4.

    2. Welded diaphragm. Stainless steel diaphragms may be welded directly to a stainless steel upper housing. The advantage of this construction is that a leak-tight joint between the two parts is readily obtained, and the lower housing can be separated from the upper

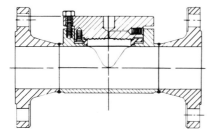

Fig. 5.3 Diaphragm seal — flow-through: type RN. (Courtesy of Mansfield & Green Division, AMETEK, Inc.)

Diaphragm Seals

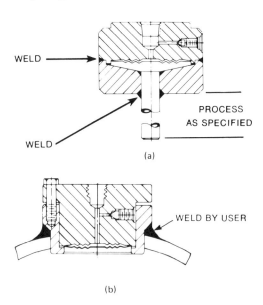

Fig. 5.4 Diaphragm seal — weld attached. (a) Type SW, (b) type SJ. (Courtesy of Mansfield & Green Division, AMETEK, Inc.)

housing without losing the fill fluid. Separation of the housings may be desirable when cleaning the seal as it permits ready access to the surface of the diaphragm. A diaphragm seal having this feature is called a "clean-out" type since the diaphragm surface and the inner surface of the lower housing can be exposed for easy removal of sludge or solidified process materials. The disadvantage in welding is that the diaphragm is not replaceable in the event it is damaged or shows evidence of corrosive attack. In this case the entire upper housing must be replaced.

3. Clamped diaphragm. Some metallic diaphragm materials and, of course, the various elastomeric materials cannot be welded to the upper housing. When using these materials the seal is made by clamping the diaphragm to the upper housing. Clamping may be accomplished between the upper and lower housings, as shown in Fig. 5.1, or by use of an auxiliary ring as shown in Fig. 5.5. Use of an auxiliary ring permits separating the upper and lower housings without losing the fill fluid. The addition of the auxiliary ring adds to the cost, so that a clean-out style diaphragm seal should be used only when necessary.

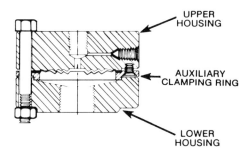

Fig. 5.5 Diaphragm seal — clean-out construction: type RA. (Courtesy of Mansfield & Green Division, AMETEK, Inc.)

4. Diaphragm protection. The inner face of the upper housing is usually contoured so that it closely matches the corrugated surface of the diaphragm. The purpose of this design is to protect the thin diaphragm against rupture if the fill fluid is lost through leakage. If this should occur, the diaphragm will be forced against, and supported by, the upper housing without damage to the diaphragm. It also permits the gauge to be removed from the seal for repair or recalibration without shutting down the process. Reinstallation of the gauge while the process system is under pressure is complicated by the fact that the refilling process must create a pressure approximately equal to the process pressure in order to force the diaphragm away from the upper housing.

5. Filling port. Some diaphragm seals incorporate a sealed filling port, which provides access to the filled volume. This is for use during the filling process (see Secs. 7.8.3 and 7.8.4). It also may be used to relieve pressure in the filled system resulting from expansion of the fluid at an elevated temperature. If, after the gauge and assembled diaphragm seal have stabilized at the process temperature and before pressure is applied, there is an indication of pressure on the gauge, the screw sealing the filling port may be loosened so as to allow the excess liquid to be expelled. The sealing screw is then securely retightened.

6. Flushing connection. Diaphragm seals may be obtained with a "flushing connection" on the lower housing (see Fig. 5.6). By means of this port the wetted surfaces of the lower housing and diaphragm can be flushed with steam or a solvent to remove sludge or other deposits without removing the entire diaphragm seal from the process.

Diaphragm Seals

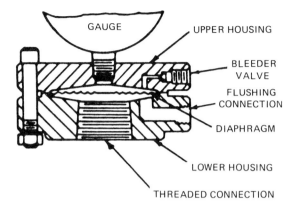

Fig. 5.6 Diaphragm seal with flushing connection.

5.2.7 Pressure Rating

Pressure rating is a matter of design and material. Thread attached, nonflow-through seals having metal lower housings are generally rated at 2500 psi at 100°F. Seals with Teflon diaphragms are limited to 1000 psi at 400°F. ASA flange-attached and inline flow-through seals will be limited to a lower-rated pressure by the pipe or flange rating. Special diaphragm seals with a pressure rating to 10,000 psi are available. Consult the manufacturer for detailed information on pressure ratings.

5.2.8 Temperature Rating

Temperature ratings of diaphragm seals depend on the fill fluid, the diaphragm material, and the lower housing material. Standard fill fluids are available for operation at temperature ranges from -60°F to +600°F. Metallic diaphragms are usable up to the limits of the fill fluid, but elastomeric diaphragms require use at lower temperatures. For example, Viton diaphragms should be used within the temperature range of -20°F to 400°F. Teflon diaphragms may be used up to 400°F. Diaphragm seals having plastic lower housings are also limited in their usable temperature range, generally 0°F to 140°F. It is advisable to consult the manufacturer with respect to use at temperatures outside these limits.

In determining the temperature limitation of diaphragm seals, it should be noted that the temperature of the process fluid is not necessarily the temperature to which the diaphragm seal is subjected. The temperature of the diaphragm seal will generally be closer to the ambient temperature than that of the process fluid by an amount that depends on the heat transfer characteristics of the particular installation.

5.2.9 Accuracy

Use of a diaphragm seal may result in some loss of accuracy with respect to the basic gauge accuracy, particularly when metallic diaphragms are used with low-pressure gauges. The more compliant elastomeric diaphragms have less effect on the accuracy. The errors attributable to stiffness of the diaphragm are usually range errors in the minus direction; that is, there is no error at zero pressure, but the indication is increasingly minus as the pressure is increased. Errors due to a pressure increase (or decrease) resulting from changes in the volume of the liquid with temperature are in the nature of a zero shift and will be more significant when they occur in conjunction with low-pressure gauges.

5.3 CASE PRESSURE RELIEF DEVICES

5.3.1 Objectives of Case Pressure Relief

As previously noted, it is always essential that consideration be given to the consequences of a ruptured pressure element. One aspect of this problem is how the pressure that is then released to the interior of the case will be vented so as not to build up sufficiently to rupture the enclosure. In general, the enclosure must have enough venting capacity so that with full-scale pressure of the intended pressure medium applied, the enclosure will remain intact even if the pressure element is ruptured. To meet this criterion, two steps are normally taken. One is to restrict the rate of flow after the rupture by installing a check or restrictor (see Sec. 5.4.4.1) in the entry port of the gauge. The other is to provide the enclosure with openings large enough to handle the flow without excessive case pressure buildup.

5.3.2 Case Pressure Relief Considerations

It is important to recognize that the check will limit the flow of fluid (gas or liquid) into the gauge, but it has no effect on the fluid that has already passed through it and is contained within the pressure element. Thus, in a high-pressure gas application, a large quantity of gas (relative to the volume of the enclosure) is stored within the pressure element. A sudden rupture of the element will release the stored gas to the interior of the enclosure at such a high rate that venting may not occur rapidly enough to prevent some portion of the closure, usually the window, from detaching or rupturing. Therefore, it must not be assumed that because an enclosure has some venting capability there cannot be a rupture of the enclosure. Such a sudden rupture of the pressure element may have the following causes: overpressuring the gauge to its burst pressure, an explosive chemical reaction within the pressurized system, weakening of the pressure element due to long-term corrosive attack, or fatigue failure. The consequences of a sudden pressure element failure are generally less severe when the gauge is used in hydraulic applications, because the fluid is essentially incompressible (except at very high pressure) and the only pressure buildup within the enclosure will be that caused by any air initially entrapped in the pressure system. It is obvious that it is more difficult to provide adequate venting of the enclosure for high-pressure gas service than for other types of service, and therefore such applications must be carefully considered.

5.3.3 Means for Providing Case Pressure Relief

For low-connected gauges, the simplest venting means is to provide case clearance around the socket. Since the socket is most often at the 6 o'clock location, adequate moisture drainage is inherent and falling dust or dirt will not enter the case. Similar case openings can be provided on low-back- and center-back connected gauges, but a drainhole at 6 o'clock may be necessary if the gauge is subject to water spray or rain.

Large openings are generally objectionable unless covered or closed in some way. Loose-fitting rubber grommets may be used to cover large-diameter holes located on the back or the periphery of the case (see Fig. 5.7). Ejection of the grommet due to internal case pressure buildup must not be restricted by the mounting arrangement or by painting it shut.

Another common method is to cover the venting holes with a thin, flexible cover that will easily deflect outward if the case pressure increases. This arrangement permits relatively large and irregularly shaped vent holes and is often used for high-pressure gas gauges. such as those used on gas welding regulators (see Fig. 5.8).

Another venting method is a plastic window having a thinned-down section adjacent to the periphery that will tear away rather than burst as a result of internal case pressure (see Sec. 3.16.1.4 and Fig. 3.61). This method is useful where a watertight case is required, because the plastic window can be sealed to the case.

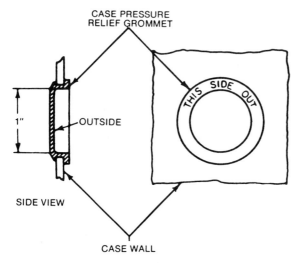

Fig. 5.7 Pressure-relief grommet.

A fourth method, and the one that provides the maximum venting area, is a pressure-relief back usually used in conjunction with solid front cases (see Sec. 3.15.9 and Fig. 3.56).

Venting liquid-filled gauges presents special problems, which are discussed in Secs. 4.3 and 4.5.

It is worth repeating that venting arrangements do not guarantee the integrity of the enclosure where sudden rupture of the pressure element occurs.

5.4 PULSATION DAMPERS (PRESSURE SNUBBERS)

5.4.1 Purpose of Pulsation Dampers

Rapidly pulsating pressure and short-term pressure surges will produce abnormal wear on the movement bearings and gear teeth and rapidly destroy the accuracy of the gauge. In addition, the indicating pointer may oscillate so rapidly that taking an accurate reading is impossible and, if the pulsation is a large percentage of the span, the bourdon will be subject to early fatigue failure. It is the function of pulsation dampers (or dampeners) to eliminate or at least reduce these adverse effects.

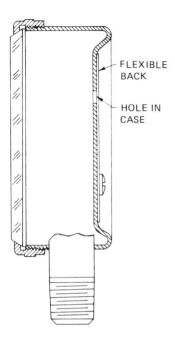

Fig. 5.8 Pressure-relief back.

5.4.2 Definition of Pulsing Pressure

Steady pressure is defined in ANSI B 40.1 (1980) as a pressure whose value varies less than 0.1% of span per second (6% per minute). Pulsating pressure varies at a higher rate. It may have a regular cycle or it may be erratic. Studies made on pulsating pressures (such as created by closing solenoid valves or gear pumps) using a high-response-rate piezoelectric sensor and an oscilloscope indicator show the frequent generation of pressure spikes several times the magnitude of the pressure indicated by the gauge. These spikes are of such short duration that the gauge mechanism cannot respond and therefore the gauge does not indicate them. Nevertheless, they persist long enough to create stresses in the bourdon, which may lead to early fatigue failure and thus loss of the pressure-containing envelope. It is therefore important to protect the gauge from pulsating pressure.

5.4.3 Pulsation Damper Considerations

Pulsation dampers are available in many forms, but all operate on the principle of restricting the rate of fluid flow into the pressure element and therefore the rate at which the pressure can change. The effectiveness of the damper is therefore a function of the size of the opening or openings in the damping device, the viscosity of the fluid, the pressure differential across the damper, and the quantity of fluid that must flow through the openings to create a pressure change within the bourdon. Restricting the flow will, of course, cause an increase in response time of the gauge (that is, the time required to indicate a pressure change), so providing damping in excess of that required is usually not desirable.

When pulsation dampers are used in hydraulic applications, it is important to note that the damping effect is dependent on the amount of air trapped in the bourdon and the entry ports of the gauge. If no air is trapped, then the amount of liquid required to change the gauge indication from zero to full scale (considering the liquid to be incompressible) is only an amount equal to the change in the internal volume of the bourdon from zero to full scale. Since there is only a very small change in the internal volume of the bourdon over the full-scale pressure range (see Sec. 3.4.2), use of pulsation dampers in hydraulic service generally requires that some air be trapped in the bourdon in order to obtain an adequate flow. Because the gauge constitutes a dead-end passage, air will normally be trapped as liquid pressure is applied. It should be noted that after a long period of usage, particularly at high pressure, some of the trapped air may be absorbed into the liquid, thus changing the damping effect of the device. If this occurs, the gauge can be removed from the process and the liquid extracted by shaking and/or evacuating the gauge; the gauge is then reinstalled. Also, it should be noted that less liquid will flow through the check in high-pressure applications than in low pressure for a given percentage of pressure fluctuation. This is because the initial volume of the trapped air is greatly reduced at the higher pressure. Therefore, there is less volume change in the trapped air (consequently less flow of the process fluid) when the pressure is increased from, say, 1000 to 1100 psi than when the pressure is increased from 100 to 110 psi. For this reason, it is more difficult to dampen high-pressure pulsations. When pulsation dampers are used in gas service, sufficient flow is generally present to provide adequate damping. However, because the viscosity of the gas is much lower than that of liquids, small passages in the damper are necessary.

Pulsation Dampers (Pressure Snubbers)

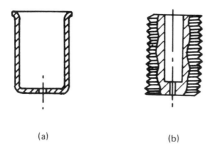

Fig. 5.9 Checks. (a) Push check, (b) threaded check.

Dampers having small passages can become clogged with foreign material in the pressure medium, preventing the gauge from indicating the proper pressure. The clogging can either prevent the entry of the medium, thus causing a low or zero indication when in fact pressure does exist, or prevent the medium from venting from the gauge port, thus causing an indication of pressure when in fact it is a lower value or zero. Use of a diaphragm seal that contains a damper operating in the liquid fill system will eliminate the possibility that the pressure medium will clog the damper.

5.4.4 Types of Pulsation Dampers

1. Checks. The simplest damper is a "check" having a small-diameter hole that is installed in the pressure port of the gauge. The check may be either a "push check" or a "screw check" (see Fig. 5.9). Push checks are the least expensive and are so called because they are pushed into the pressure port and retained by an interference fit. The screw check can be easily removed for cleaning or replacement or to permit removal of fluid from the gauge. The checks are usually made from the same material as the socket. The hole size will vary with the application. A hole 0.007 in. in diameter is often used in high-pressure gas service, and a hole 0.022 in. in diameter may be used for liquid service. Generally, the gauge manufacturer will install a push check or screw check in the gauge if specified by the buyer. High-pressure gauges are usually furnished with a check even if not specified, in order to limit the rate of flow of the pressure medium in the event the pressure element develops a leak.

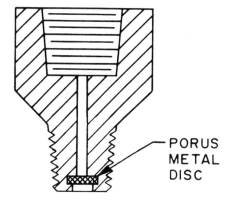

Fig. 5.10 Porous metal pulsation damper.

2. Porous metal. Another type of damper is made using a porous metal disc as the flow restrictor. This type of damper is usually supplied in a separate housing or fitting that is interposed between the gauge and the pressure source rather than being built into the gauge (see Fig. 5.10). Discs having a wide range of porosity are available and are usually made of stainless steel. The housing is available in various materials. Greater flow restriction can be obtained with this type damper than with a check.

3. Energy absorber. Another type of damper utilizes one or more small-diameter rods operating in cylinders having a carefully controlled clearance. Flow is limited in accordance with the magnitude of the clearance. The reciprocating action of the rods, resulting from the pressure pulsations, tends to make this type self-cleaning. As with the porous metal dampers, this type is supplied in a separate fitting rather than built into the gauge.

4. Needle valves. Where the magnitude and frequency of the pressure pulsations and surges vary so that no single damper is suitable, a needle valve (see Fig. 5.11) can be installed between the gauge and the pressure source. Under actual operating conditions, the valve is adjusted to dampen out the undesirable pointer fluctuations. As conditions change, readjustment can be easily made. For severe conditions, the valve offers the possibility of completely closing off the pressure source to the gauge except when it is necessary to take a reading, at which time it can be

Pulsation Dampers (Pressure Snubbers)

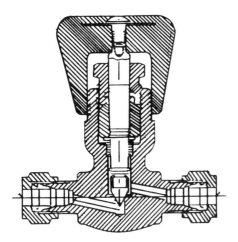

Fig. 5.11 Needle valve.

opened very slightly (cracked) and reclosed. Clogging is minimized since the valve can be fully opened momentarily in order to clear blockage around the valve seat. Where such an arrangement can be used, the gauge life will be greatly extended.

5. Surge tank. For severe pressure pulsations it may be necessary to install a surge tank in the pressure line at some point near the gauge. The surge tank provides a volume that will tend to even out rapid pressure changes since sufficient fluid must enter or leave the tank to equalize the line pressure. By selecting the proper size tank, the time necessary to equilize will be long compared to the frequency of the pulse.

6. Capillary. Frequently, pressure pulsation is accompanied by mechanical vibration that also adversely affects the life of the gauge. Connecting the gauge to the pressure source by means of a long length of capillary having, say, a 0.020-in. bore not only will dampen out the pressure pulsations but will also permit mounting the gauge on a nonvibrating surface or to a shock-mounted panel. The capillary may be coiled into a helical form in order to save space.

7. Check valve. Dampers of the type described above will afford an indication of the average pressure (that is, the pressure midway between the lowest and highest applied), assuming that

the pressure extremes persist for the same time period. If it is desired to obtain an indication of the maximum pressure to which the gauge is subjected, a check valve can be installed in the pressure line that will readily permit flow into the gauge and restrict flow out but not prevent it completely. This can be done by scoring the valve seat or by providing a restricted bypass line around the check valve. Pressure peaks of extremely short duration cannot be measured by mechanical devices because of the high inertia of the various parts. Solid-state pressure transducers with an oscilloscope readout are used for such measurements.

5.5 GAUGE COCKS

Gauge cocks are rotary plug valves that are installed in the pipeline leading to the pressure gauge. The cock provides a means of shutting off the line so that the gauge can be removed from the line without loss of pressure. The valve handle shows whether the valve port is open or closed, being parallel to the valve body when it is open. Fig. 5.12 illustrates a common type that is available in a 1/4-in. NPT size and with pressure ratings of 150 to 250 psi steam (300 to 600 psi hydraulic). The same care is necessary in selecting the valve material and pressure rating as in the selection of the gauge itself, for it will be exposed to the same measured fluid. As mentioned in Sec. 5.4.4.4, a gauge cock can also serve as a pulsation dampener; conversely, a needle valve installed for pulsation damping can also serve as a shut-off valve.

5.6 SIPHONS

5.6.1 Definition and Purpose

The term "siphon", as applied to devices described in this section, has been in existence for perhaps a century. Nevertheless, it is a misnomer because the devices do not provide any siphoning action. Siphons are most often used when measuring steam pressure, although they may be used in any system containing condensible vapors. Saturated steam at a pressure of 100 psi is at a temperature of 337°F, which is well above the 120°F recommended maximum exposure temperature for gauges having pressure elements with soft soldered joints. The purpose of the siphon is to provide, in effect, a heat exchanger where the steam can be condensed and the resulting condensate cooled prior to entering the pressure element. In addition to protecting the soft soldered joints of the

Siphons

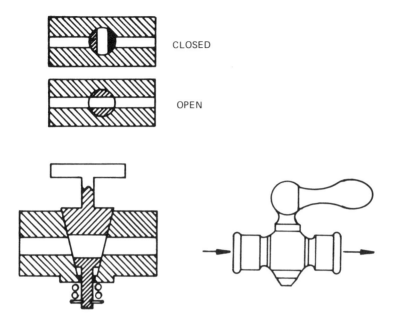

Fig. 5.12 Gauge cock.

pressure element, the reduction in temperature of the pressure medium reduces the error of indication that results from the change in spring rate (modulus) of the bourdon material with temperature. Generally, gauges having silver brazed or welded pressure elements will not require a siphon for the protection of the pressure element. However, it may be desirable to provide a siphon for the purpose of keeping the pressure element closer to the temperature at which it was calibrated. This is especially true for high-pressure steam boilers where the temperature of super heated steam may be sufficiently high to anneal some copper alloys. The material of the siphon will vary with the process on which it is installed, but in general it should be the equivalent of the pressure element. As a general guide, Fig. 5.13 lists the maximum pressure and temperature limits for siphons made from several materials. Be certain to check with the manufacturer to determine if a particular siphon will be suitable for the application before installing the siphon.

Size (in.)	Material	Pressure and temperature limits
1/4	Iron	500 psi and 400°F
1/4	Brass	250 psi and 400°F
1/4	Extra heavy seamless steel	1000 psi and 850°F
1/2	Extra heavy seamless steel	1000 psi and 850°F

Fig. 5.13 Pressure and temperature limits.

5.6.2 Types of Siphons

1. Pig tails. One form of a siphon is called a pig tail and consists of a length of tubing coiled as shown in Fig. 5.14. This form provides a large cooling surface and a trap that prevents the condensate from draining away, thus effecting a liquid seal through which incoming vapor must pass. Because of this feature, siphons are sometimes referred to as traps.

2. Bulb siphons, internal siphons. Siphons come in forms other than the pig tail illustrated. These include bulb siphons, wherein a spherical chamber is used to trap the condensate, and internal siphons. The latter are similar in principle to the pig tail, but the coil is contained within the gauge case, usually in the form of a helix coiled into a diameter larger than that of the bourdon. Since siphons trap the condensate, the possibility of freezing during periods of nonuse must be considered. With the ready availability of brazed or welded pressure elements, the use of siphons is infrequent.

5.7 BLEEDERS

If the pressure element of a gauge must be internally flushed, filled with liquid, or drained of liquid, a bleeder arrangement can be incorporated to facilitate the operation.

Basically, the bleeder provides a port in addition to the normal entry port so that the pressure element is not a dead-ended volume. Bleeders are particularly useful when a bourdon is the pressure element and usually are positioned at the tip end (see Fig. 5.15). With

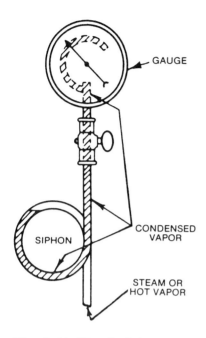

Fig. 5.14 Pigtail siphon.

the bleeder open, liquid can be flushed through the bourdon in order to facilitate the cleaning operation. When making measurements of liquid pressure, the liquid level will rise within the bourdon as the pressure is increased and the air trapped within the bourdon is compressed. The pressure of this liquid column will be added or subtracted from the applied pressure (see Sec. 6.4.2 for a more detailed explanation). For precise measurements this may be a significant factor. The error due to this variation in liquid head can be eliminated by completely filling the bourdon with the pressure medium. In this manner the error becomes a constant value and can be calibrated out. A bleeder tip makes the filling operation relatively simple because it permits the pressure medium to be flushed through the bourdon until no air exits from the bleeder port, at which time it is sealed.

If the bourdon contains undesirable liquid, such as might accumulate as a condensate from compressed air containing oil and/or water vapor, then it can be removed by opening the bleeder and

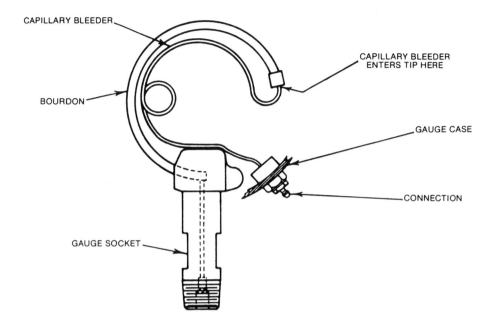

Fig. 5.15 Bleeder tip with external connection.

blowing out the liquid with air pressure, followed by flushing with a solvent. The form illustrated in Fig. 5.15 is the most convenient since the bleeder port is located outside the case and the effluent can be easily collected. A less expensive form consists of only a length of capillary closed by crimping. When it is necessary to use the bleeder, the crimped portion is cut off and subsequently recrimped. The simplest form has a threaded hole in the tip that can be closed by some sealing means, such as a set screw and ball arrangement. Use of either of the latter forms requires partial disassembly of the gauge, but this may be acceptable for occasional use. Entrapped air within the bourdon will accumulate at the highest point in the system, so when purging the air the orientation of the bourdon must be changed during the process to be certain the air finds its way to the bleeder port.

Another form of bleeder is illustrated in Fig. 5.16. This arrangement contains a flexible plastic tube within the bourdon that extends to the highest point of the assembly. The other end exits

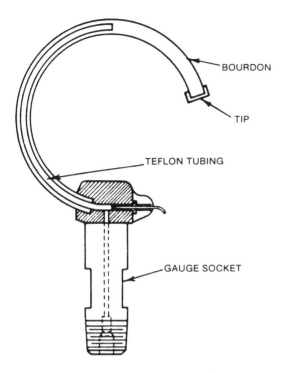

Fig. 5.16 Bleeder—internal capillary.

from the socket as shown. With this construction it is not necessary to reorient the gauge during purging.

5.8 HEATERS FOR GAUGES AND CONNECTING LINES

For applications where the surrounding temperature can become low enough to freeze the measured fluid inside the gauge element or the connecting piping, some form of heater must be installed. If the gauge is close to the line or vessel whose pressure is being measured, and the measured medium is a higher temperature, it may be necessary only to insulate the connecting line to the gauge, as shown in Fig. 5.17. This arrangement will prevent heat loss to the surrounding cold air and thus prevent freezing of the measured

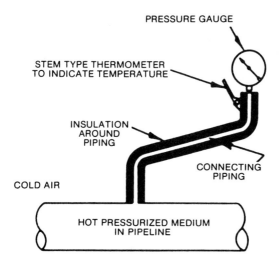

Fig. 5.17 Steam-traced gauge.

fluid in the connecting line and the gauge itself. If additional heating is required to prevent freezing within the gauge and its connecting lines, or if the temperature must be raised to reduce the viscosity of the medium, the gauge and connecting lines can be "steam-traced." This simply means that suitable tubing is coiled around or fastened adjacent to the areas to be heated; insulation is applied to retain the heat. Steam is then passed through the coil to provide the necessary heat. A typical application of steam tracing is shown in Fig. 5.18. Electric heaters are commercially available that closely fit standard gauge cases and may be more convenient and more readily installed than steam lines. Whichever type of heater is used, care must be taken so that the maximum operating temperature of the gauge is not exceeded.

5.9 MAXIMUM AND MINIMUM POINTERS

5.9.1 Purpose

It is sometimes desirable to indicate the maximum and/or minimum pressure attained during some event. For example, in a hydraulically operated tensile testing machine, the specimen under test is

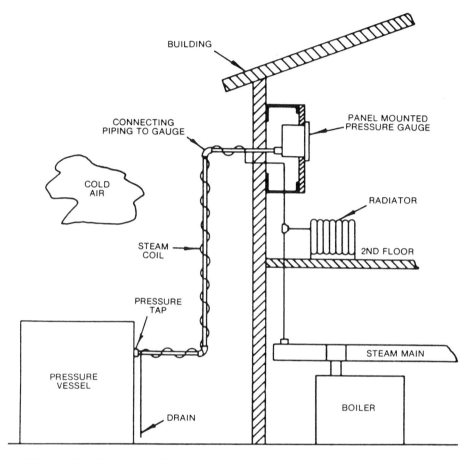

Fig. 5.18 Steam-traced system.

pulled until it fractures. The hydraulic pressure at the instant of fracture is a measure of the force needed to break the specimen. However, when the specimen breaks, the pressure is suddenly released and the exact pressure indication at that instant is lost. By using a maximum pointer, the highest pressure applied during the test will be indicated and a reading readily made. Another use for maximum and minimum pointers (often called "lazy hands" or "telltale pointers") is to indicate the pressure extremes over some period

158   Chap. 5 Gauge Accessories

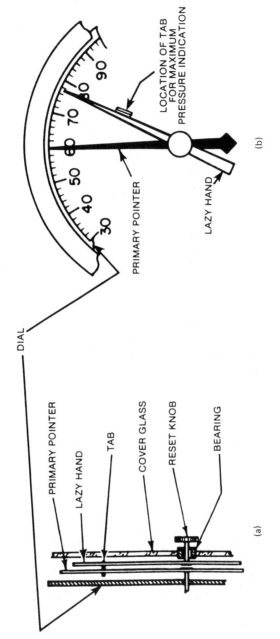

Fig. 5.19 Max–min pointer. (a) Cross section, (b) front view.

of time without the necessity of having an observer constantly watch the process or of installing a pressure recorder.

5.9.2 Construction

Max—min pointers are usually mounted on the inside of the gauge window so that they can rotate concentrically with the primary pointer, which is driven by the pressure element. A small tab on the lazy hand is picked up by the primary pointer, causing the lazy hand to rotate with the primary pointer. If the lazy hand is intended to indicate the maximum pressure, then the tab is arranged to contact the up-scale edge of the primary pointer. If the lazy hand is for indication of a minimum pressure, the tab contacts the down-scale edge of the primary pointer (see Fig. 5.19). If both maximum and minimum indications are needed, then, of course, two lazy hands are required, one having an up-scale tab and the other a down-scale tab. A reset knob is usually provided so that after a reading is taken the lazy hand can be reset against the primary pointer for the next event.

5.9.3 Max—Min Pointer Considerations

One of the problems with lazy hands is that they retard the rotation of the primary indicating pointer due to friction in the lazy hand pivot. The magnitude of the retardation can be determined by allowing the primary pointer to push the lazy hand to some point, carefully observing the primary pointer indication, and then moving the lazy hand out of the way and observing the change in indication of the primary pointer. From this standpoint it would appear that the pivot should be as frictionless as possible. However, it is necessary to provide some means by which the lazy hand can retain its position until it is reset. Further, there must be sufficient friction to keep the lazy hand from coasting to a spurious position if it is hit sharply by the primary pointer. The best ways to minimize these problems are to balance the lazy hand about its rotational axis so that it has no tendency to rotate under vibration, keep the lazy hand assembly as light as possible so it will have a minimum inertia, and apply a film of silicone oil or grease to the pivots to provide viscous damping. Silicone fluids have a relatively constant viscosity over a wide temperature range and like all viscous materials will resist rapid motion but yield to a steady pressure. For critical applications it is sometimes necessary to customize the lazy hand by varying the pivot clearance and the viscosity of the silicone so as to take into account the force available at the primary pointer, the rate of the pressure change to be indicated, and the vibratory conditions to which the gauge is subjected. Vibration is a factor, as it tends to make the lazy hand rotate if it is not properly balanced.

5.10 ELECTRICAL SWITCHES

5.10.1 Application of Switches

When the measured pressure reaches a predetermined maximum or minimum value, it is sometimes desirable to turn on a signal light, sound an alarm, or operate a pump or valve. These actions can be accomplished by providing the gauge with one or more sets of electrical contacts so that a circuit can be closed or opened at the desired pressure. There are on the market many varieties of pressure switches, that is, devices that open or close contacts at a preset pressure but that do not also indicate the pressure. These are termed "blind" switches. In this discussion we will cover only the indicating-type pressure switches.

5.10.2 Abutting Contacts

Simply incorporating a set of abutting contacts (see Fig. 5.20) that can be operated by some moving part of the gauge (i.e., bourdon, segment, or pointer) has limited applicability for the following reasons:

1. Open contacts are easily contaminated and are subject to physical damage.
2. Once the contact is made, further motion of the indicating mechanism is severely retarded. Use of sliding contacts, wherein a wiper is made to slide across a commutator containing conductive segments, will reduce the retarding effect if the wiper pressure is kept low but will create a drag on the bourdon over the entire range of the contact closure.
3. If the pressure changes slowly, the contacts will, at some pressure near the operating point, be sufficiently close to cause arcing and burning of the contacts, particularly when opening. Also, if vibration and/or pulsation is present, the contacts may be subjected to rapid make and break again, resulting in arcing and burning. Under these conditions it is possible for the contacts to fuse together, giving a permanent "make" signal, or oxidize so that no make signal can be given.

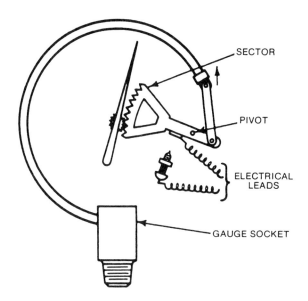

Fig. 5.20 Alarm circuit—abutting contacts.

4. Contacts of this type can handle only a very low current because the contact pressure necessary to provide a good electrical circuit is not compatible with the low operating force of the pressure element.

5.10.3 Snap Action Switches

Some of the above objections may be overcome by replacing the abutting contacts with commercially available snap action switches. In these, the electrical contacts snap from an open to a closed position (or vice versa), providing a substantial separation of the contacts for a small amount of motion at the actuating point. The contacts are contained within a plastic housing giving them some protection from corrosion and mechanical damage. However, snap action switches require considerable force (relative to that available from the bourdon) to operate, and will retard the indication while the snap action mechanism is being cocked prior to actual contact movement. This will cause the indication of the pressure at the switching point to be in error. Continued motion of the bourdon after the

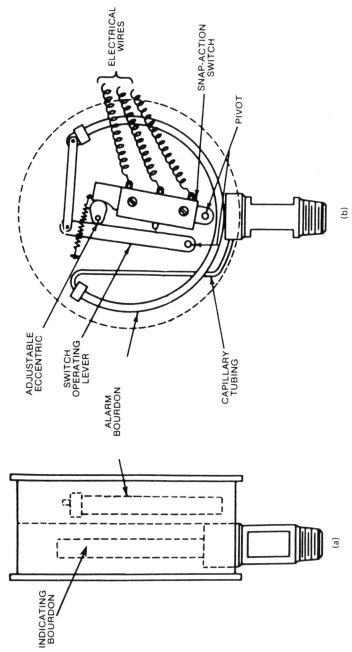

Fig. 5.21 Indicating pressure switch. (a) Side view, (b) rear view.

switch has operated will be more severely retarded, or even stopped entirely, once the switch has "bottomed."

Snap action switches have a built in "switching differential," which means that the "make" point occurs at a point in the travel of the switch plunger other than the "break" point. Therefore, if the switch contacts make at, say, 100 psi on increasing pressure, it may be necessary to reduce the pressure to 95 psi to break the contacts. Vibration performance will be improved over that of abutting contacts, but arcing, burning, and chattering may still occur, particularly if the pressure changes very slowly, because the contact pressure reduces to zero at the break point. The current and voltage rating of indicating switches must be carefully considered, and it is advisable to discuss the requirements with the manufacturer. Resistor-capacitor networks and various solid-state components are available which will reduce arcing, or a time delay circuit can be employed that will maintain a closed circuit even if the contacts are chattering. If the current-carrying capacity of the switches is not large enough to handle the device to be controlled (for example, a large heater load), the switches can be arranged to control an auxiliary relay that does have sufficient current-carrying capacity.

5.10.4 Explosive Hazard

Pressure switches of the type described above must not be used where there is a possibility of the gauge operating in an explosive atmosphere because arcing between contacts provides a source of ignition.

5.10.5 Use of Dual Bourdons

Retardation of the indication can be overcome by incorporating two bourdons in parallel, using one bourdon for the indication and the other for switch operation. In Fig. 5.21 a single snap action switch is positioned for operation by a bourdon and linkage mechanism. A second bourdon operates the pointer in the usual manner. As shown, the switch plunger will be depressed and activate the switch on increasing pressure. If a separate alarm is required to be actuated at a lower pressure, then a second snap action switch may be positioned on the opposite side of the switch-operating lever so as to depress the switch plunger on decreasing pressure.

5.10.6 Circuit Design

Snap action switches can be obtained with a "common" terminal (C) and both a "normally open" (NO) and a "normally closed" (NC) terminal. Such a switch is called a single pole double throw (SPDT) since it will switch an incoming circuit to either one of two outgoing circuits. In operation, when the switch plunger is in its normal position (i.e., not depressed by the operating lever), there will be a closed circuit between the common terminal (C) and the normally closed (NC) terminal. There will be an open circuit between the common terminal and the normally open (NO) terminal. When the plunger is depressed sufficiently to operate the switch, then there is a closed circuit between the C and the NO terminals and an open circuit between the C and the NC terminals. If, for example, it is desired to light a lamp when some predetermined pressure is reached on ascending pressure, then the lamp would be wired in series with a power source and the C and NO terminals. As the pressure increases, the switch plunger will be depressed so as to actuate the switch, the C and NO contacts will make, and the lamp will turn on. Alternatively, if a pump motor is to be shut down when a predetermined pressure is reached, then the motor starter relay would be connected in series with the C and NC terminals so that, as the pressure increased and the switch operated, the circuit between the C and NC terminals would be opened. It is possible, of course, to utilize all three terminals so that when the switch operates, a lamp will light and the motor will stop. Operation of a switch on descending pressure follows a similar sequence.

5.10.7 Circuit for High—Low Limit Switching

It is often desired to control the pressure between both a high and a low limit; for example, to start a pump when the pressure falls to some minimum value and allow it to run until the pressure reaches the desired maximum value. As stated earlier, a snap action switch has a built-in differential between the open and closed positions. However, this differential cannot be controlled by the user of the switch and in any case is generally not large enough to provide the desired difference between the open and closed switch positions. By using an auxiliary relay wired as shown in Fig. 5.22, both the high and low operating pressure points can be controlled. Referring to Fig. 5.22, as the pressure falls, the switch operating lever will actuate the low switch closing the C and NO terminals. This will actuate the electromagnetic coil of the auxiliary relay, closing the relay contacts 1 and 2 and starting the pump motor. As the pressure increases, the operating lever moves to the left and at some point will release the low switch plunger and break the circuit between the C and NO terminals. Normally, this would deactuate the relay. However, current will continue to flow through the relay coil by

Electrical Switches 165

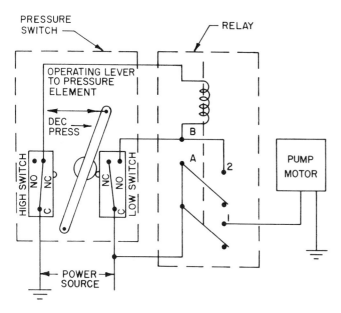

Fig. 5.22 Pump switch circuit.

reason of the now closed relay contacts and the circuit between points A to B. The motor will therefore continue to run even though the low switch is no longer actuated. As the pressure continue to increase, the high switch will be actuated, opening the circuit between the C and NC terminals. When this occurs, the relay will be deactuated and the motor will stop. In this manner, both the high and low operating points can be controlled as the switches may be individually positioned.

5.10.8 Magnetic Reed Switches

Use of magnetic reed switches will largely eliminate the retarding effect of the snap action switches because the reed switches require very little force to operate. Magnetic reed switches are illustrated in Fig. 5.23 and are actuated when a small magnet is positioned adjacent to the switch. The actuating magnet is small enough to be attached to the indicating pointer of a pressure gauge so that reed switches may be positioned around the periphery of the dial. The switches can be quite small and several can be accommodated if

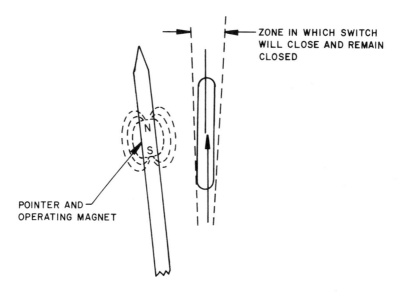

Fig. 5.23 Reed switch.

desired, although usually only a high and a low switch are used. As the indicating pointer and magnet assembly approaches the switch (for example, on increasing pressure), the magnetic field of the magnet will actuate the switch and the contacts will close. As the pressure continues to increase, the magnet will move away from the switch and the contacts will open. If, as is the general case, it is desired to maintain the contacts in the closed position for all pressures higher than the actuation point, a bias magnet can be added. The bias magnet is permanently fastened on the reed switch in such a position that its magnetic field will be strong enough to maintain the contacts closed once the switch is actuated, but not strong enough to actuate the switch if the contacts are open. The polarities of the bias magnet and the pointer magnet are oriented with respect to each other so that as the pointer moves upscale it will reach a point where the magnetic field between them intensifies the field surrounding the soft iron reeds of the switch, and the contacts will close, after which the field of the bias magnet is sufficiently strong to maintain the contacts in the closed position even though the pointer magnet continues upscale. On decreasing pressure, a point

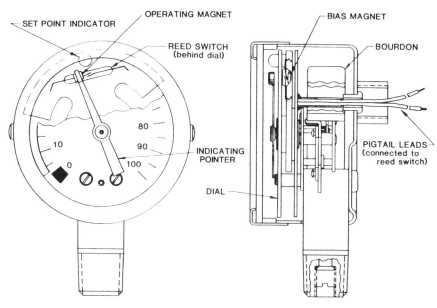

Fig. 5.24 Gauge equipped with reed switch.

is reached where the field holding the two reeds together is shunted away from the reeds by the field between the pointer and bias magnets, and the contacts will open. Because of the magnetic hysteresis of the system, the deactuating (drop-out) position of the pointer will be downscale from the actuating (pull-in) position, resulting in the switch-opening pressure being less than the switch-closing pressure. If the magnets are properly designed, this switching differential can be kept quite small, typically 2% of the pressure span of the gauge. Fig. 5.24 illustrates the manner in which reed switches can be incorporated into a pressure gauge.

5.11 SHUT-OFF VALVES AND GAUGE PROTECTORS

Pressure gauges may be protected from damage from overpressure by use of a shut-off valve in the line leading to the pressure gauge. Shut-off valves are designed to close at a preset pressure and to reopen when the pressure falls below the preset value. This is accomplished by means of a spring-loaded piston that holds a valve stem away from its valve seat. The applied pressure acts on the piston so that at some pressure the force of the spring will be overcome and the piston will move and allow the valve stem to seat, cutting off any further increase in the applied pressure. Valves of this type are commercially available in various materials. Shut-off valves may be used where it is desirable to indicate pressure to an accuracy based on percent of indicated reading. For example, if a 0–10,000 psi gauge having an accuracy of ±1/2% is used to indicate the pressure in a system, the accuracy of any reading would be ±50 psi. This may not be acceptable at an indication of 1000 psi, where the accuracy would be ±5% of the 1000 psi. By using one or more shut-off valves, a series of gauges having increasing values of span can be utilized. Thus, four gauges in spans of 0–300 psi, 0–1000 psi, 0–3000 psi, and 0–10,000 psi will give good readability and an indication having an accuracy of at least 2% of indicated reading at all pressures above 75 psi as shown in the chart.

Range of gauge (psi)	Accuracy of gauge (psi)	Range of operation (psi)	Accuracy at lowest reading (%)
0–300	±1.5	75–250	±2.0
0–1000	±5.0	250–750	±2.0
0–3000	±15.0	750–2500	±2.0
0–10000	±50.0	2500–7500	±2.0

Of course, the gauges of 0–300, 0–1000, and 0–3000 psi must be protected by shut-off valves. In applying shut-off valves, the response time of the valve must be compatible with the rate of pressure rise. That is, the valve requires time to close, and rapidly peaking pressure pulses and shock waves may simply pass through the valve.

6
Gauge Selection and Installation

6.1 INTRODUCTION

The factors that are normally considered for the selection and installation of a pressure gauge are:

1. Nature of the pressurized medium, including its required pressure range
2. Environmental conditions
3. Method of connecting gauge to pressurized medium
4. Method of mounting gauge
5. Required size of gauge
6. Required accuracy of gauge

Each of these factors is considered in the following paragraphs, and, equipped with this and the other information on gauge components and accessories contained in other chapters of this handbook, the user should be able to select a gauge for a specific application that will operate properly with minimum maintenance.

6.2 NATURE OF PRESSURIZED MEDIUM

6.2.1 Pressurized Medium Considerations

Because the measuring element (elastic chamber) of a pressure gauge is exposed directly to the pressurized medium, needless troubles can be avoided by being knowledgeable about the nature of the medium. It may be corrosive, it may solidify at normal atmospheric temperatures, or it may contain tars or solids that will leave deposits inside the elastic chamber. Such conditions can impair the performance of the gauge or make it completely inoperative.

6.2.2 Chemical Compatibility

The use of pressure gauges for the measurement of strong oxidizing fluids — for example, oxygen, hydrogen peroxide, and chlorine — requires special consideration. These and similar materials will react violently with many organic materials, especially when pressurized. Such a violent reaction can almost instantaneously raise the pressure within the bourdon to a value far exceeding the burst strength of the bourdon. If this occurs the bourdon will fragment, releasing high-pressure gases to the interior of the enclosure, which may cause the enclosure to fail, usually by blowing out the window. Severe injury and/or property damage can result. Therefore, it is extremely important for the user to specify that the gauge is to be used for this type of service (see Sec. 2.6.3) so that the manufacturer will control the level of internal contamination during manufacture. Also, it is extremely important that the user maintain the cleanliness of the gauge. For example, if the gauge is checked or recalibrated prior to installation, pressurization must be done with oil-free air or water, and fittings must be free from grease and oil. Most compressors impart a small amount of oil to the air or water as it is being pumped, and this must be removed to insure gauge cleanliness.

The pressure-containing envelope must be corrosion-resistant with respect to the pressurized medium. Published tables giving the corrosion rates of various chemicals and materials may be used as a guide in making a selection. However, it must be recognized that bourdons operate with highly stressed, relatively thin walls. The accuracy of indication is directly affected by any reduction in the wall thickness such as may result from corrosion. A material having a corrosion rate of 0.001 in. per year may be suitable for piping or tanks, but it will not be satisfactory for a bourdon having a wall thickness of, for example, only 0.010 in. Further, the rate of corrosion of bourdons is often accelerated due to the necessarily high internal stresses developed when pressurized. It is best to be conservative when selecting the material of the pressure-containing element for corrosion resistance. In order to provide some degree of standardization, the great majority of pressure gauges have pressure-containing elements made from either a copper alloy (brass,

Nature of Pressurized Medium

phosphor bronze, or beryllium copper), stainless steel (type 316 or 403), or Monel. Other materials are generally considered specials with commensurate high cost and long delivery time. If none of the standard materials is suitable, it may be less expensive to buy a standard gauge assembled to a diaphragm seal (see Sec. 5.2) in order to obtain the required corrosion resistance.

6.2.3 Temperature Compatibility

Temperature of the pressurized medium must be considered. Most pressure gauges, particularly those made with copper alloy bourdons, have soft soldered joints in the pressure containing envelope. The shear strength of a 50—50 soft solder (50% tin and 50% lead) at 250°F is only one-quarter of the strength at 85°F. Further, soft solder will creep under a sustained load at high temperature and the joint will fail. If the temperature of the pressurized medium can cause the temperature of the pressure-containing envelope to exceed approximately 120°F, then the gauge should be isolated (see Sec. 5.6) or gauges having silver brazed or welded joints should be specified. Heating of the pressure-sensitive element will cause a loss of accuracy due to dimensional changes of the internals and to the change in the modulus of the element material (see Secs. 6.3.2 and 6.3.3 for further discussion).

6.2.4 Magnitude of Pressure

The magnitude of the pressure will determine the pressure range selected. The American National Standard Institute's publication B40.1-1980, *Gauges — Pressure Indicating Dial Type — Elastic Element*, recommends that the range of a pressure gauge be selected so that the operating pressure occurs in the middle half of the scale, that is, between 25 and 75% of the span. As a guide, the full-scale pressure of the gauge should be twice the operating pressure. By doing so, fatigue life will be improved and a margin provided in the event the operating pressure exceeds its intended value. Occasional application of pressure up to the maximum range of the gauge will not be detrimental.

It is also important to know if the pressure to be applied to the gauge is steady or pulsating. Sec. 5.4 discusses the effects of pulsating pressure and the ways in which they can be minimized.

When selecting the span of the gauge, it is of course wise to select one of the manufacturer's catalog standards in order to avoid the long delivery time and the excessive cost associated with "specials." The ANSI B40.1 standard referred to in Sec. 2.1 lists the

Positive pressure

in. H_2O/psi

in. H_2O		psi	psi	psi
0/10	in. H_2O	0/3	0/200	0/10 000
0/15		0/5	0/300	0/15 000
0/30		0/10	0/600	0/30 000
0/60		0/15	0/1000	0/60 000
0/100		0/30	0/1500	0/100 000
0/200		0/60	0/3000	
0/300		0/100	0/6000	

kPa

0/1	0/10	0/100	0/1000	0/10 000	0/100 000
0/1.6	0/16	0/160	0/1600	0/16 000	0/160 000
0/2.5	0/25	0/250	0/2500	0/25 000	0/250 000
0/4	0/40	0/400	0/4000	0/40 000	0/400 000
0/6	0/60	0/600	0/6000	0/60 000	0/600 000

bar

0/0.01	0/0.10	0/1.0	0/10	0/100	0/1000
0/0.016	0/0.16	0/1.6	0/16	0/160	0/1600
0/0.025	0/0.25	0/2.5	0/25	0/250	0/2500
0/0.04	0/0.40	0/4.0	0/40	0/400	0/4000
0/0.06	0/0.60	0/6.0	0/60	0/600	0/6000

Negative pressure (vacuum)

in. Hg	kPa	bar
-30/0 in. Hg	-100/0	-1.0

Compound pressure

in. Hg/psi	kPa	bar
30 in. Hg vac/15 psi	-100/150	-1/1.5
30 in. Hg vac/30 psi	-100/300	-1/3
30 in. Hg vac/60 psi	-100/500	-1/5
30 in. Hg vac/100 psi	-100/900	-1/9
30 in. Hg vac/150 psi	-100/1500	-1/15
	-100/2400	-1/24

Fig. 6.1 Preferred ranges.

	Numeral interval			Value of smallest interval		
Span	2A	A	B	2A	A	B
0–15	1	3	3	0.1	0.2	0.5
0–30	5	5	5	0.2	0.5	1
0–60	5	10	10	0.5	1	2
0–100	10	10	20	1	1	2
0–200	20	20	40	2	2	5
0–300	20	50	50	2	5	5
0–600	50	100	100	5	10	10
0–1000	100	100	200	10	10	20

Fig. 6.2 Typical numeral and graduation intervals (all values in psi).

preferred spans shown in Fig. 6.1. It will be noted that, with few exceptions, gauges using psi or in. H_2O as the unit of pressure have spans of 1, 1.5, 3, or 6 multiplied by 10^n and gauges using kPa or bar as the unit of pressure have spans of 1, 1.6, 2.5, 4, or 6 multiplied by 10^n. The value of n can be zero or any negative or positive whole number. Other spans are often cataloged, particularly for ranges less than 15 psi where in. Hg and oz/in.^{2} are used as the pressure unit.

The frequency of numerals and major graduations and the value of the smallest graduation interval are less standardized. The B40.1 standard does not recommend specific values for these parameters but does recommend that they be determined using intervals of 1, 2, or 5 x 10^n, where n is as defined above. For example, if the span is 300 psi, the interval between major graduations could be either 10, 20, or 50 psi, but not 25 or 30 psi. If the interval between major graduations is 20 psi, the value of the interval between minor graduations may be 2 psi (10 spaces) or 5 psi (4 spaces), but not 4 psi (5 spaces). Depending on the size of the dial and required nomenclature, it may not be possible to completely follow these recommendations.

Figure 6.2 tabulates some typical spans, the numeral interval, and the value of the smallest interval between graduations in accuracy grades 2A, A, and B.

6.3 ENVIRONMENTAL CONDITIONS

6.3.1 Environmental Considerations

There are many environmental conditions that will affect the accuracy and life of a gauge. These include temperature, vibration, mechanical shock, humidity, corrosive atmospheres, and dust.

6.3.2 Effect of Temperature—Range Shift

One of the most important environmental considerations is the effect of ambient temperature. In addition to possible deterioration of the soft soldered joints in the pressure-containing envelope, the accuracy may be adversely affected. This loss of accuracy is the result of two distinct physical changes. The first of these is a change in the modulus of elasticity of the bourdon (change in spring rate). The change in spring rate causes a "range shift," resulting in errors which increase proportionally as the pressure is increased. The error resulting from the spring rate change may be as much as 2% of the indicated reading for each 100°F increment of temperature change. Therefore, a gauge with a span of 0–100 psi that was calibrated at 75°F might have added errors at a temperature of 175°F equal to +2 psi at 100 psi indication and +1 psi at 50 psi indication. If the initial accuracy of the gauge was 1% of full scale, then the errors at 175°F might be 3 psi at 100 psi indication and 2 psi at 50 psi indication.

6.3.3 Effect of Temperature—Zero Shift

Temperature extremes introduce an additional error called "zero shift," which is caused by the physical change in the dimensions in the various parts of the gauge with temperature. The magnitude and direction of this error vary with gauges of different designs and are functions of the relative thermal expansion of all of the various parts of the gauge, involving materials, methods of attachment, and dimensions. Zero shift is a constant shift over the entire scale regardless of whether or not pressure is applied to the gauge, and the magnitude of the shift does not vary with applied pressure as it does with range shift.

6.3.4 Compensation for Range Shift

There are two basic methods for overcoming range shift due to temperature changes. One is to use a bourdon material that does not change spring rate with temperature. There are several such

Environmental Conditions 175

materials available, but the one with the best combination of physical properties is Ni-Span C. This material is hardenable by heat treatment, and by varying the heat-treating temperature the change in the elastic modulus, called the thermoelastic coefficient, can be maintained at zero, positive, or negative. This material will maintain the elastic modulus over a usable temperature range of approximately -50°F to +150°F. With temperatures in excess of these values, a small change takes place. However, Ni-Span C bourdons are very expensive to manufacture.

The second method for correcting range shift consists of using bimetallic linkage elements arranged to increase or decrease the range of the instrument as the ambient temperature decreases or increases. This is done by varying the multiplication ratio of the movement. This method is more critical to manufacture and less stable than the method of using a material having a constant elastic modulus. It also has the disadvantage that if a maximum stop is used to limit pointer travel during applied overpressure, the relatively weak bimetallic link may be distorted. A further disadvantage in the use of bimetallic temperature compensation for spring rate is that the motion of the bi-metal, in correcting for range, may also impart a slight change in the linearity of the indication.

6.3.5 Compensation for Zero Shift

Zero shift is usually corrected by using some type of bimetallic link [see Sec. 3.12.6.4(e)]. The same disadvantages as cited above for bimetallic range corrections are in order. Refer to Sec. 3.12.6.4(f) for discussion of another type of multimetallic linkage mechanism that can be adjusted to give the required amount of correction to compensate for zero shift.

6.3.6 Maximum Temperature Limits

The maximum ambient temperature normally considered practical, even with temperature compensation, is 250°F. At such high temperatures, special dial materials must be used, and soft soldered joints are not acceptable. In extremely cold environments, movement pivots and gears may bind. Where there is danger of the pressure medium freezing, heaters should be used (see Sec. 5.8). Remember that the actual temperature of the measured liquid or gas in a pipeline or vessel can be much higher or lower than the ambient temperature at the gauge. When considering the need for temperature compensation, the determining factor is the actual temperature of the gauge, not the temperature of the measured liquid or gas.

6.3.7 Mechanical Shock and Vibration

Most gauges will not withstand repeated mechanical shocks and continuous vibration severe enough to cause the pointer to oscillate. If excessive vibration is encountered, it may be possible to mount the gauge at another location and connect the gauge to the pressure source with rubber or plastic hose or a length of coiled metal tubing. If this is not possible, the manufacturer can supply the gauge with a movement having a high viscosity silicone compound applied to the teeth and/or pivots. The silicone will aid in damping the pointer oscillation and impart some lubrication to the moving parts. For severe vibration the use of liquid-filled gauges may be the best solution (see Chap. 4).

6.3.8 Humidity and Corrosive Atmospheres

Pressure gauges will continue to operate satisfactorily in a humid atmosphere and normal outdoor environments if the enclosure is plastic, a copper alloy, or stainless steel and the movement is of similar materials. However, commercial-class gauges utilizing steel cases, rings, and dials will show some deterioration in outdoor applications and will have to be replaced periodically. Plastic cases withstand outdoor exposure very well and are a less expensive alternative to stainless steel cases. For more severe atmospheres, as might be encountered in close proximity to acid baths, a fully corrosion-resistant gauge, consisting of a stainless steel case and a stainless steel or plastic movement and dial, may be necessary. Dust is not usually detrimental except under very severe conditions, in which case the solution is to provide a sealed enclosure with some means of venting. For extremely severe environments, liquid-filled gauges having an all-type 316 stainless steel construction should be considered.

6.4 METHOD OF CONNECTING GAUGE TO PRESSURIZED MEDIUM

6.4.1 Strength Considerations

Gauges are almost always supplied with pressure connections having male pipe threads 1/8, 1/4, or 1/2 NPT (American National Pipe Taper). Gauges having other types of connections are available but generally are not stocked, so long delivery and setup charges may be encountered. The use of commercially available adapters is an alternative to employing gauges with special connections. Gauges in ranges over 20,000 psi may be supplied with special

Method of Connecting Gauge to Pressurized Medium 177

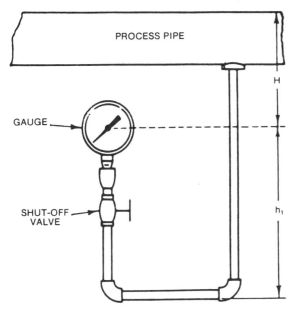

Fig. 6.3 Head effect — gauge below connection.

high-pressure connections; unless the user has experience in such high-pressure systems he should discuss such applications with the manufacturer. Gauges having ranges above 10,000 psi are normally furnished with steel or stainless steel sockets. For pressures of 25,000 psi and higher, high-tensile materials must be used. It is generally recommended, although not mandatory, that gauges of 4 1/2-in. and larger should have 1/2 NPT connections because of their weight and greater susceptibility to being broken off when struck.

6.4.2 Correction for Liquid Head

For precise measurement, it may be necessary to correct the gauge after installation to nullify the effect of a liquid head pressure. As shown in Fig. 6.3, a gauge located below the pressure tap has a water column head H acting on the gauge (the head h_1 is on both legs of the connection and cancels out). As discussed in Chap. 1 of this handbook, such liquid heads can be converted to psi equivalents. The significance of such heads is greater with lower-pressure spans. For example, in round figures, if H is 5 ft (Fig. 6.3), it will add 2

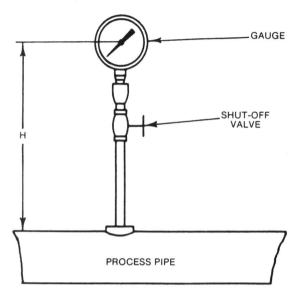

Fig. 6.4 Head effect — gauge above connection.

psi to the measured pressure. If the gauge span is 500 psi, this represents an error of 0.4%; with a span of 100 psi, the error increases to 2%; and with a span of 15 psi, the error becomes 13%. If the gauge is installed above the pressure tap, as shown in Fig. 6.4, the head H subtracts from the measured pressure and a positive zero correction must be made.

Where the installation involves a liquid head acting on the gauge that must be compensated, be sure that the connecting piping is completely filled with liquid. This is accomplished with, for example, a shut-off valve at the pressure source, thereby enabling the connecting line to be filled prior to gauge installation. When the connecting line is completely filled, the shut-off valve is closed and the gauge may now be installed. To compensate for the liquid head, measure the effective height and calculate the equivalent psi (plus or minus) to obtain the appropriate correction.

6.4.3 Installation

When installing pressure gauges, always use the wrench flat provided on the pressure connection to tighten the gauge into the mating fitting. Never apply torque to the gauge case. It is generally necessary to use a suitable thread sealant, preferably a compound rather than a tape. Tape shreds and can get into the pressure port of the gauge, causing a blockage. Gauges having dampers or checks are particularly susceptible to blockage in this way.

6.4.4 Effect of Leakage

Leakage in the connecting line to a pressure gauge means that there will be some flow of the pressurized medium. Depending on the magnitude of the leak and the restriction to flow in the connecting line, there will be a pressure drop between the source of the pressure and the leak, and the gauge will indicate a pressure less than the pressure at the source. Therefore, lines leading to a pressure gauge and the connection of the gauge in the line must be leak-free for accurate measurement.

6.5 METHOD OF MOUNTING GAUGE

6.5.1 Introduction

Various mounting arrangements are discussed in Sec. 2.4.2, and the bolt circle and panel cutout and the diameter of the mounting bolt-hole dimensions for panel-mounted gauges are given in Fig. 3.46. It is important that, regardless of how the gauge is mounted, the case vent or vents be free to operate in order to avoid excessive internal case pressure should a leak develop in the pressurized components. Gauges may be stem-mounted, provided the supporting connection is rigid and not subject to vibration. Outside case dimensions, that is, depth, major diameter, connection length, etc., will vary among different manufacturers and among different case styles from the same manufacturer. Therefore, if the dimensions are critical, request a manufacturer's certified outline drawing, which will give the basic dimensions and their expected tolerances.

6.5.2 Orientation

Gauges are usually designed for installation in a vertical position, that is, with the midscale point at 12 o'clock and the dial in a vertical plane. This is the position in which the mechanism is designed to operate and also the position in which the gauge is calibrated. In other positions, the bearing loads in the movement are changed and the overhung weight of the cantilevered bourdon acts in a direction different from that in which the gauge was calibrated. The resultant shift in calibration, called "position error," will usually be larger on low-pressure gauges because the bourdon is less rigid. Position error is essentially a constant and can be corrected by adjusting the pointer. Gauges can be constructed to give trouble-free operation in positions other than the normal vertical position. If other than vertical mounting is required, specify the exact mounting position so that the manufacturer can provide the proper construction for this service.

6.5.3 Provision for Tapping Prior to Reading

The rated accuracy of a pressure gauge is based on a reading taken after lightly tapping the gauge in order to overcome the effect of friction between the moving parts. If the mounting is in a location not readily accessible for tapping and the accuracy of the indication is critical, a small electric buzzer can be mounted on the gauge case or connecting piping. Just prior to reading, the buzzer may be activated from a remote location. A buzzer has a secondary advantage in that it applies a constant level of vibration, whereas tapping by hand will vary from one person to another.

6.5.4 Special Mounting Methods

Special mounting methods and connecting means (including designing into the user's enclosure) can be provided, particularly when large quantities are required. In such instances the gauge manufacturer will be able to make specific suggestions and recommendations.

6.6 REQUIRED SIZE OF GAUGE

Choosing the right size of gauge requires consideration of the required accuracy and readability. As noted in Sec. 2.7.2, higher-accuracy gauges are necessarily larger in order to provide sufficient scale length to resolve the degree of accuracy. A gauge having a nominal diameter of 2-in. may be obtained with an accuracy of 1%. However, it is difficult to read because the 1% represents only 0.04 in. of motion at the tip of the pointer. Reading discrepancies of at least 1% will occur, particularly if the pointer is between graduations, and time will be lost in making observations. If 1% accuracy is really needed, it would be cheaper in the long run to buy a 4 1/2-in. gauge. Another reason for choosing a larger gauge is to provide better readability when the gauge is observed from a distance.

For this reason, gauges having only a grade B accuracy (see Sec. 2.7.6) are available up to 4 1/2-in. in diameter.

6.7 REQUIRED ACCURACY OF GAUGE

Accuracy of pressure gauges is discussed in Sec. 2.7. As noted, increased accuracy means increased initial cost as well as increased cost to insure maintenance of the required accuracy. Therefore, it is best to specify only the grade of accuracy that is actually required. In severe service, particularly under conditions of pulsing pressure and vibration, it is very difficult to maintain a high degree of accuracy. In such situations, it is better to sacrifice accuracy in order to get either a more rugged gauge, or a gauge in a lower accuracy classification which can be economically replaced when the wear becomes excessive.

Since variations in the ambient temperature affect the accuracy of a gauge, it is necessary to use it at the temperature at which it was calibrated if the stated accuracy is to be realized. This may not be possible, in which case the additional cost of the more accurate gauge is difficult to justify. Remember, the rated accuracy of a gauge is based on its being read after lightly tapping the gauge in order to remove friction in the moving parts.

7
Gauge Maintenance and Calibration

7.1 INTRODUCTION

After a pressure gauge is put into service, it will be necessary to check periodically to determine that it is functioning properly, and that it is in adequate condition to continue to perform its intended use. The purpose of this chapter is to offer practical guides in testing, calibrating, and maintaining pressure gauges.

Before starting to calibrate a gauge, check the rated accuracy as stated by the manufacturer. If, for example, the accuracy of the gauge, as supplied, was ±2% over the middle half of the scale, it may be difficult or impossible to calibrate it to ±0.5%.

7.2 CONDITIONS AFFECTING ACCURACY AND PERFORMANCE

Many conditions will adversely affect the accuracy and overall useful life of a pressure gauge, the most detrimental being pulsating pressure, vibration, and internal and external corrosion. All of these topics are discussed in detail elsewhere in this handbook. The important point here is that any maintenance program must be based on the type of service and environment in which the gauge is operating. The frequency of calibration necessary to insure that the gauge is operating satisfactorily and within the required accuracy will vary with the operating conditions, the required accuracy, and the consequences of a false indication of pressure. It is generally more economical to replace gauges in the commercial class (see Sec. 2.5.1) than to repair and/or recalibrate them. Even so, a maintenance program should cover these gauges as well as more expensive types to insure that they are functioning in the intended manner. Many users find it most economical to replace commercial gauges on a periodic basis, which is established by experience with the particular application, rather than wait until the gauge becomes inoperative.

7.3 MAINTENANCE PROGRAM

The purpose of a gauge maintenance program is to preserve the useful life and continued accuracy of indication of the gauges used in the facility. Proper maintenance begins with proper selection of the gauge for the intended use. It follows also that the gauge must be properly installed. It will be necessary to periodically repair and service the gauges in use. If serious damage has occurred, it will be necessary to replace the gauge.

Failure of the pressure element is usually obvious. The cause(s) of the failure should be determined so that the situation can be corrected. Causes of failure of the pressure element are often corrosion, overpressure, and fatigue. Corrective measures for failures of these types include proper material selection, use of a gauge having a higher pressure span, and use of snubbing or damping devices.

As part of a maintenance program, records should be kept on the usage and service requirements for the gauges used by the facility. This will include calibration records, records and analyses of failures, and such other information as is required for control of stocked gauges, parts, costs, and sources of supply.

7.4 MAINTENANCE FACILITY

7.4.1 General Considerations

A complete maintenance facility will include the necessary equipment for calibrating pressure gauges. Calibration is generally defined as (1) comparing the pressure indicated by the gauge being calibrated with the pressure indicated by a pressure standard, and (2) making adjustments to the gauge being calibrated so as to make the indication agree with the standard as closely as required. Some people interchange the words calibration and testing. Others differentiate the two by defining calibration or calibration verification as the process of determining the errors of a gauge, and testing as the process of making adjustments to reduce the errors. In this text the general definition will be used. Calibration cannot be accomplished by simply setting the pointer on zero at zero pressure. To begin with, many commercial class gauges do not show a zero graduation, and second, setting the pointer on zero does not allow for errors in range or scale shape that may be inherent in the gauge. Further, the highest accuracy is usually desired at about midscale rather than zero. A controlled pressure source and pressure standards are therefore mandatory if calibration is to be accomplished.

7.4.2 Pressure Source

1. Air pressure. A pressurized air supply can be obtained from either a compressor or, if the use is infrequent, a tank of compressed air. If it is important to keep the pressure-containing envelope free of contamination, the use of tanks of dry nitrogen is recommended. Either of these sources represents a convenient method of calibrating gauges up to about 600 psi. However, several important cautions are in order.

If a compressor is used, the air should be filtered to remove any oil or water. This is very important where contamination of the internal surfaces of the pressure-containing envelope must be avoided, such as in gauges for oxygen service.

Be sure to use a pressure regulator to reduce the pressure source to a value only slightly in excess of the span of the gauge being tested. It is virtually impossible to test a 100-psi gauge that is connected to a 1000-psi pressure source through a simple valve. It is also dangerous, since inadvertent opening of the source valve can apply sufficient pressure to burst the bourdon.

When calibrating any pressure gauge, and particularly when calibrating with air or other compressed gas, it is wise to provide a protective transparent shield between the operator and the gauge. This is especially important when calibrating used gauges and gauges whose history is unknown and which may burst as a result of having a defect in the pressure-containing envelope due to corrosion or fatigue.

2. Vacuum. For testing vacuum and compound gauges it is necessary to have a vacuum pump capable of creating an absolute pressure of 0.1 psia or less.

3. Hydraulic pressure. For pressure above 600 psi and up to 15,000 psi it is safer to use a hydraulic source. A deadweight tester is the most common means of providing hydraulic pressure

for calibrating purposes, since it provides not only the pressure source but also a very accurate measurement of the pressure. A discussion of the theory of deadweight testers is given in Sec. 7.5.6. When using a hydraulic deadweight tester, is is inevitable that some of the hydraulic fluid will get into the gauge. The contamination thus incurred may be unacceptable, particularly if the hydraulic fluid is oil. For this reason it may be best to obtain a deadweight tester using water as the fluid. It is suggested that the objectives of the particular maintenance shop be discussed with the manufacturer of the deadweight tester and his recommendations followed as to the equipment which should be used.

It is unlikely that it will be necessary to generate pressure above 15,000 psi, so a hydraulic deadweight tester will be sufficient. Working with pressures above 15,000 psi requires special equipment and techniques and is best left to the manufacturer of the device to be tested.

7.4.3 Pressure Standards

In addition to controlled pressure sources, a maintenance shop will require a suitable array of pressure standards. There is a variety of standards that may be used, including deadweight testers (both hydraulic and pneumatic), precision test gauges, and manometers. Tradeoffs between precision, cost, convenience, and objectives of the maintenance shop must be made in selecting the standards. Various standards are reviewed in Sec. 7.5.

7.4.4 Miscellaneous Equipment

Various small tools, including a pointer puller (Fig. 3.39) and tapered reamers for reaming pointer bushings, will be needed, as well as cleaning solvents and polishing paper. A supply of spare parts such as windows, movements, and pointers, and a storage area for parts salvaged from gauges considered unrepairable will also be of value.

7.5 PRESSURE STANDARDS

Pressure standards may be categorized as (1) primary standards, (2) secondary standards, and (3) line standards. Alternative terms are, respectively, (1) reference standards, (2) transfer standards, and (3) measuring and test equipment.

7.5.1 Primary Standards

Primary standards are the most accurate of the three types. They establish the basic accuracy values for the calibration system, and their principle of operation is based on the fundamental units of mass, length, and time. Primary standards are used to calibrate secondary standards. Certificates should be available for all primary standards. The certificate shows the calibration date, accuracy, the environmental conditions under which the accuracy is derived, and a statement explaining the traceability to the National Bureau of Standards. Primary standards are certified by higher-echelon laboratories of nationally recognized companies with respect to calibration and traceability to the National Bureau of Standards.

7.5.2 Secondary Standards

Secondary standards are also instruments of high accuracy but are of lesser accuracy than primary standards and can tolerate normal handling as opposed to the extreme care usually necessary with primary standards. Secondary standards are used as a medium for transferring the basic value of primary standards to lower-echelon or line standards, and are checked against primary standards.

7.5.3 Line Standards

Line standards are all devices and equipment used to measure, test, inspect, or otherwise examine items to determine compliance with specifications, and are checked against secondary standards.

7.5.4 Required Accuracy of Standard

It is desirable, of course, to have the accuracy of the calibrating apparatus as high as possible when compared to the gauge being calibrated. Ten times the accuracy of the gauge under calibration is considered ideal. However, when checking high-precision test gauges this may not be possible, and standards having accuracies two to three times the accuracy of the gauge under calibration are acceptable.

7.5.5 Application of Standards

1. During test or calibration, pressure must be applied to the gauge under test. At the same time, through a tee connection (see

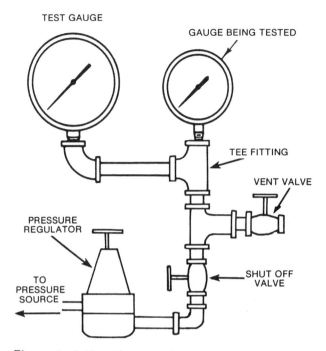

Fig. 7.1 Calibration stand.

Fig. 7.1), pressure is also applied to a standard of known accuracy. The readings obtained from the standard and the gauge under test are compared to determine the errors.

2. Standards should be connected close to the gauge being calibrated and must be carefully handled in order to maintain accuracy. Connecting lines must be leak-tight (see Sec. 6.4.4).

3. All standards should be identified as to their class (e.g., line standard), and a program for periodically checking them against the next higher standard should be established. Consultation with the gauge manufacturer will be helpful in establishing a proper schedule.

7.5.6 Deadweight Testers

1. Attainable accuracy. Deadweight testers are frequently used as primary standards, since they may be obtained to an accuracy as

high as 0.03% of indicated pressure with data furnished certified traceable to the National Bureau of Standards. With good care they can maintain their accuracy over a long period of time.

2. Theory of operation. A deadweight tester operates on the principle of supporting a known weight (force) by means of a pressure acting over a known area, thus fulfilling the definition of a primary standard as being based on mass, length, and time. The weights for a given tester are normally identified in terms of pressure (rather than weight). Fig. 7.2 shows the construction of a typical hydraulic deadweight tester.

3. Correction for gravity. In order to obtain the high precision, many factors and corrections must be considered by the manufacturer. Simply dividing the weight by the area of the piston will not be sufficient (see Sec. 7.5.6.7). The force produced by the loading weights used with the deadweight tester is a result of the attraction of the earth's gravitational field on the mass of the loading weight. Therefore, the actual force created will vary at different locations on the earth because the magnitude of the gravitational field varies at different locations. The variation in the gravitational field is about 0.5% (maximum to minimum) of the total field. The variation in local gravity from that at which the deadweight tester was manufactured may be sufficiently large to require compensation. For this reason, higher-accuracy testers are furnished "corrected to local gravity," which means that the weights are made heavier or lighter than the nominal weights so that they will create the same force in the locale where they are used that the nominal weight would at the standard gravity of 980.665 cm/sec^2.

Limitations due to decreasing sensitivity, resulting primarily from friction, make it difficult to maintain accuracy at the low end of the range. Generally, deadweight testers are used over the upper 90% of the rated maximum pressure.

4. Pneumatic deadweight testers. Pneumatic deadweight testers are available for the measurement of pressure in the ranges of 4 in. H_2O to 300 psi and 10 psi to 1000 psi. One unique type uses a ceramic ball instead of a piston. A schematic diagram of this pneumatic tester is shown in Fig. 7.3. In this construction, a precision ceramic ball is floated within a tapered stainless steel nozzle. A flow regulator introduces air under the ball, lifting it and forming an annulus between the ball and nozzle. Equilibrium is reached when the vented flow equals the fixed flow from the supply regulator, at which time the ball floats. The pressure thus established, which is also the output pressure, is proportional to the weight load. During operation the ball is centered by a dynamic film of air, eliminating physical contact between the ball and nozzle.

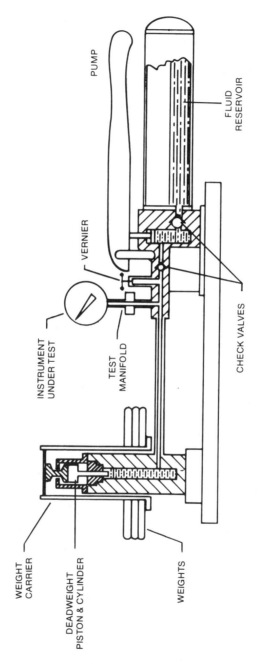

Fig. 7.2 Hydraulic deadweight tester. (Courtesy of Mansfield & Green Division, AMETEK, Inc.)

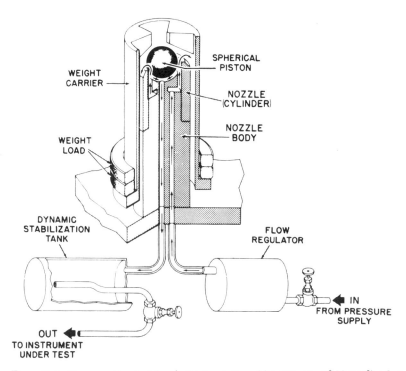

Fig. 7.3 Pneumatic deadweight tester. (Courtesy of Mansfield & Green Division, AMETEK, Inc.)

When weights are added or removed from the weight carrier, the airflow changes. The regulator senses the change in flow and adjusts the pressure under the ball to bring the system into equilibrium, changing the output pressure accordingly. Thus, regulation of output pressure is automatic with changes in weight on the spherical piston (ball). Since the ball floats on a layer of air, friction is virtually nonexistent within the device. This results in excellent repeatability, and hysteresis is virtually eliminated. A further advantage is the elimination of the necessity of weight rotation. The floating ball piston is self-cleaning while in operation and does not require cleaning as do conventional pistons and cylinders. These testers are completely self-regulated and the accuracy is virtually independent of any operator techniques.

5. Hydraulic-pneumatic deadweight testers. Combination hydraulic-pneumatic deadweight testers are available wherein a separating device is connected into the test stand so that air pressure

can be applied to the gauge under test and also to the hydraulic fluid of the deadweight tester. The air pressure is then carefully regulated to raise the piston of the deadweight tester to the point at which the gauge can be read. Extreme care is required when operating such a system since there is a danger of blowing out the piston. Further, any use of high-pressure gas is potentially dangerous because of the compressibility of the gas.

6. Operation of hydraulic deadweight tester. To operate the deadweight tester, the oil level is first checked and replenished if necessary. The gauge should be attached so as to minimize the effect of oil head pressure. With medium- and high-pressure gauges it should be sufficient to mount the gauge so that the bottom of the case is at the same level as the bottom of the piston (remember that the piston moves approximately 1/2 in. off its rest position under test conditions). Since hydraulic deadweight testers are not normally used for testing low-pressure ranges, it is not within the scope of this discussion to outline the procedures to correct for oil head pressures when calibrating low-pressure-range gauges. The weight for the desired calibrating pressure is now placed on the carrier. Note that the weight of the piston assembly must be included in the total weight. The pump handle is operated to pressurize the system until the piston is freely floating midway between the piston rest and the overpressure arrest. Some manufacturers engrave a line on the piston to indicate the correct position of the carrier. When the correct position has been reached, the piston and weights are slowly and continuously rotated while the indication of the gauge under calibration is observed.

Pressure may be varied by adding or removing weights on the carrier. However, this necessitates repositioning of the piston. It may be necessary to open the relief valve, depending on the size of the weight removed, to avoid raising the piston too high. After a test, release the pressure by opening the relief valve. Do not disconnect the gauge until the pressure is completely vented.

The weights of a deadweight tester represent definite pressure increments. These can be as low as 1-psi increments, depending on the range. Several piston sizes are usually included with the tester to extend the range. Pressure ranges up to 15,000 psi are commonly available.

7. Factors affecting accuracy. Factors affecting the accuracy of pressure measurement when using a deadweight tester as the pressure source are principally the precision of the mass of the loading weights and the effective area of the cylinder and piston. Other factors that affect the accuracy of the measurement but are not considered significant in the calibration of even the most precise bourdon

Pressure Standards 193

gauges are the effect of air buoyancy on the weights, effect on the piston of the surface tension of the hydraulic fluid, and the thermal expansion and elastic deformation of the piston and cylinder. Of even less significance are corkscrewing, weight stack vertical alignment, and contamination. Eccentric rotation of the weight stack may cause a side thrust on the piston and cylinder, which will create excessive wear and possibly inaccurate readings. It is good practice to bolt the tester to a solid bench and level it. A complete discussion on the subject of the accuracy of deadweight testers can be obtained from National Bureau of Standards Monograph No. 65.

7.5.7 Liquid Columns (Manometers)

1. General considerations. Use of liquid columns (usually mercury or water) for the measurement of pressure is based on the principle that an applied pressure will support an observable column of the liquid against the gravitational pull on the liquid. The higher the pressure, the higher the column that can be supported. The liquid is contained in a glass tube of constant bore having a graduated scale behind it so that the height of the column can be determined (see Fig. 7.4). The force created by the column and balanced by the pressure is a function of the height of the column, the density of the fluid, and the earth's gravitational field at the location. There are many factors that affect these three parameters, such as the temperature effect on the fluid and the graduated scale, the variation in gravitational force at the locale in which the measurement is being made, the surface tension of the liquid, and the cleanliness of the apparatus. The manufacturer of the liquid column will take these factors into consideration when the column is designed and calibrated, making the necessary corrections. He will furnish correction factors for conditions other than those he considers standard. A complete treatise on mercury manometers is given in the National Bureau of Standards Monograph No. 8 and, of course, the column manufacturers will furnish further information.

2. Units of pressure. As noted elsewhere, units of pressure based on liquid columns, such as inches of mercury and centimeters of water, are not recognized by the ISO standard. Only the pascal is recognized. However, liquid columns are often used for the measurement of low pressure, and as a matter of convenience the use of these units of pressure will probably continue.

3. Measuring vacuum. Mercury columns are frequently used to measure negative pressure (vacuum), in which case the negative

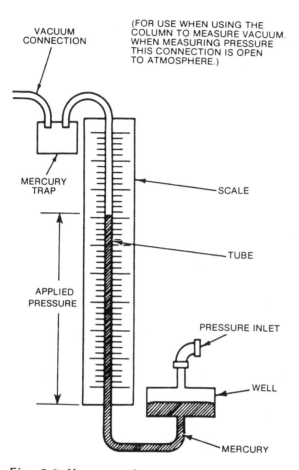

Fig. 7.4 Mercury column.

pressure is applied to the top of the column and the well is open to atmospheric pressure.

4. Safe handling of mercury. Mercury columns should be equipped with traps to catch any overflow due to improper operation. Mercury and mercury vapor are dangerous substances if they enter the body through cuts, inhalation, or ingestion. If any spillage occurs, the mercury should be cleaned up immediately using approved equipment and procedures. Never let mercury come into

Pressure Standards 195

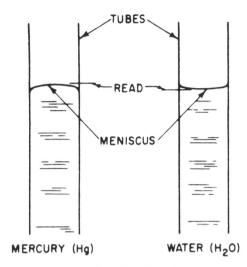

Fig. 7.5 Reading liquid columns—meniscus.

contact with parts made of brass, bronze, or other copper alloys, as it will form an amalgam that will cause failure of the part.

5. Reading liquid columns. It will be noted that in a mercury column the meniscus is convex, whereas in a water column it is concave. This is due to the different values of surface tension of the two liquids. Water "wets" the inside of the tube, indicating a low surface tension; mercury does not, indicating a high surface tension. It is customary to read the level at the center rather than at the inside wall of the column, as shown in Fig. 7.5.

7.5.8 Test Gauges

The use of precision bourdon-actuated test gauges as secondary or transfer standards is very convenient. They can be obtained having an accuracy of 0.1% of full scale, which is adequate for checking grade 2A (±1/2%) gauges and others having lesser accuracy. They may be temperature-compensated, and are usually furnished with a certified chart listing the actual error so that more precise readings can be made by applying the correction. It must be realized, however, they are more subject to wear and damage than a deadweight tester is and therefore require more frequent calibration. Utilization of precision test gauges in conjunction with periodic recalibration using a deadweight tester may be the most economical way to set up an adequate maintenance shop if the variety of gauges and the frequency of calibration are not great.

7.5.9 Pressure Transducers

A variety of digital readout pressure standards utilizing outputs from strain gauges, piezoresistive transducers, and capacitance transducers is available. They are generally expensive and require periodic checking but are very convenient to use and offer a wide range of pressure spans. Devices having accuracies of up to 0.05% are available. Chapter 9 presents an overall review of the various types. If buying such devices is contemplated, literature and prices from the manufacturers should be obtained and compared. A listing of manufacturers can be found in instrumentation trade journals or by contacting the local chapter of the ISA (Instrument Society of America).

7.6 CALIBRATION OF PRESSURE GAUGES

7.6.1 General Procedure

The gauge to be calibrated is mounted on a test stand (see Fig. 7.1) or on a deadweight tester with the dial in a vertical plane unless it is known that the actual installation is otherwise. The standard against which the gauge is to be compared is connected in parallel, and pressure is applied from one of the sources previously discussed. There should not be any leak in the system. Check by closing off the pressure source and noting if the pressure in the system is maintained. Just prior to the start of calibration, the gauge being calibrated should be subjected to a pressure equal to the full-scale pressure of the gauge. Pressure is then applied in approximately equal increments in both increasing and decreasing directions. A minimum of three test points is required in class B gauges, up to five test points on class 2A gauges, and 10 test points on class 4A gauges. The gauge under test should be lightly tapped before each reading. If precision test gauges are used as the standard, they must also be tapped lightly before reading. It is convenient to attach an electrical buzzer to the test stand as an alternative to tapping.

7.6.2 Alternative Reading Methods

Usually a precision test gauge (because of its higher accuracy) will have a greater number of graduations and increased reacability

Calibration of Pressure Gauges

compared to the gauge being tested. For this reason, it is preferable to apply sufficient pressure to bring the pointer of the gauge under test to a specific graduation and then read the value of the applied pressure on the precision test gauge, thus improving the ability of the observer to estimate the value of the applied pressure when the pointer is between graduations.

When using a deadweight tester, the smallest increment of pressure that can be applied to the gauge being tested is determined by the smallest weight in the set of weights. Therefore, it is necessary to apply a specific pressure by means of the deadweight tester and determine the error by reading the gauge under test. Summarizing, using a precision test gauge as a standard and a variable pressure supply in the manner described above results in precisely knowing the indication of the gauge under test (because the pointer is set precisely on a graduation) and estimating the applied pressure (because the pointer of the precision test gauge is likely to be between graduations). Using a deadweight tester as the standard produces the opposite condition; that is, the applied pressure is precisely known, and the indication of the gauge under test must be estimated.

This leads to the matter of applying algebraic signs when recording the difference between the indications of the gauge being calibrated and the standard. Assume a 0—100 psi gauge is being calibrated and that it actually indicates 47 psi when a true pressure of 50 psi is applied. If sufficient pressure is applied to set the pointer of the gauge being calibrated on the 50 psi graduation, the indication on the standard gauge will be 53 psi. The observer must not be confused by the fact that the standard gauge is reading 3 psi plus and thereby record the error as +3 psi. If the standard gauge is reading higher than the gauge being tested, then the error in the gauge being tested must be a minus error, and, in the example given, the error is properly recorded as -3 psi. There is less chance for confusion when using a deadweight tester since (continuing with the same example) 50 psi would be applied to the gauge and the pointer would indicate 47 psi. The observer would properly record this as a -3 psi error.

7.6.3 Error vs. Correction

It is also important to understand the difference between "error" and "correction" in order to be sure the algebraic sign is properly applied.

If E is the error, C the correction, R the reading, and T the true pressure (as determined by the standard), then the sign of an "error" in indication of a gauge is determined by the relation E = R - T and the sign of a "correction" is determined by the relation C = T - R. It also follows that E = -C.

From the above it will be seen that error is the difference between the indicated value and the true value of the applied pressure. A positive error denotes that the indicated value is greater than the true value. Correction is the quantity that is algebraically added to an indication value to obtain the true value. The algebraic sign of the correction is opposite to the sign of the error.

7.6.4 Application of Error Charts and Correction Charts

Standards are often furnished with a chart giving the error of the standard at selected test points. This can be given in the form of an error or a correction. Continuing with the same example as above, the chart furnished with the 0–100 psi gauge would record the error as -3.0 psi. This tells you that the gauge reads less than the applied pressure, so when 50.0 psi is applied it will indicate only 47.0 psi. The chart could also record the correction as +3.0 psi. This tells you that indication of the gauge is to be corrected by adding 3.0 psi, so if it reads 47.0 psi the true pressure is 50.0 psi.

7.6.5 Recording Readings

In recording readings it is inappropriate to record significant figures beyond the accuracy of the reading. For example, assume that a 0–100 psi gauge having 1-psi increments is being calibrated and at some test point it is estimated that the pointer is 1/4 of a space high. To record the error as +0.25 psi is misleading, since it implies an accuracy of measurement that is not really obtainable. Visually splitting the space between the graduations into 10 parts is the best that can be done, and therefore the observer should record either +0.2 psi or +0.3 psi, whichever is the more appropriate.

Readings should be recorded on a data sheet such as the one shown in Fig. 7.6. Charts attached to the gauge showing the error at the various test points are most often in the form of "error" rather than "correction," as there seems to be less confusion in the proper application.

7.6.6 Classification of Errors

If the errors are unacceptable, then it will be necessary to make adjustments to the gauge mechanism in order to reduce the errors.

TEST RECORD DATA SHEET

Manufacturer _____ Checking std. _____

Size – span _____ Frequency ___ Yr. ___ Mos. ___ Days

Rated accuracy _____ Location _____

Identification No. _____ Date checked _____ By _____

Pressure unit _____

Test std. pressures	Gauge reading		Error before cal.		Error after cal.		Conditions and remarks
	Up	Down	Up	Down	Up	Down	

Fig. 7.6 Test record data sheet.

The observed errors are the result of friction, hysteresis, range, scale shape, and zero shift, acting either singly or in combination. The following paragraphs describe each type of error and the corrections to be made. Frequently the corrections are interrelated, making it necessary to make some adjustments more than once to achieve the best accuracy.

7.7 CORRECTION OF ERRORS

7.7.1 Friction

1. Definition. Friction is the difference between the pointer indication before and after light tapping. Additionally, this condition may be observed by sluggish or erratic pointer movement either up or down scale. Friction may be caused by many factors, the most common of which are dirt, excessive wear, and incorrect hairspring tension. (Note: While hairspring tension is the term used in the industry, it would be more proper to refer to it as hairspring torque.)

2. Correction. When sluggish or erratic pointer action is noted, it will be necessary to clean all bearing surfaces and gear teeth. This can usually be accomplished by washing the movement in a bath of suitable industrial solvent. The unit should then be properly dried with a light blast of clean, dry air. When using volatile cleaning fluids, take proper precautions for adequate ventilation.

Should the contamination and dirt be very stubborn, it may be necessary to completely disassemble the movement to gain access to all bearing holes. Great care should be exercised so as not to damage or distort the hairspring. If excessively worn parts are evident they should be replaced at this time. Upon reassembly, the hairspring must be properly adjusted to eliminate backlash in the teeth, link, and bearings. To set the proper tension of the hairspring, it is necessary to disengage the sector and pinion. This may be accomplished by lifting the pointer over the dial stop pin or by removing the limiting zero pressure stop (if used) and manually deflecting the pressure element so as to disengage the pinion and sector gear teeth. In high-pressure gauges having very stiff bourdons this may not be possible, and it will be necessary to remove the link in order to get the sector out of mesh with the pinion. The bourdon may be permanently damaged by manually deflecting in excess of approximately 180° of the pointer travel. When in doubt, remove the link.

In some movements, it may not be possible to remove the stops so as to unmesh the segment. In this case, provision is usually made to manually deflect the sector off the end of the pinion by pushing the sector in a direction parallel to the pinion.

Having unmeshed the sector, note which direction of pointer travel causes the hairspring to wind more tightly. Usually, the maximum hairspring tension is at zero indication (hairspring most tightly coiled) and the hairspring unwinds as the pointer moves upscale. With this construction, the unmeshed pointer should be turned counterclockwise (thus winding up the hairspring) approximately 1 1/4 turns and the segment remeshed. When the pointer is at zero indication, there should be 1 to 1 1/4 turns on the hairspring. At maximum pointer travel, there should be 1/4 to 1/2 turns on the hairspring (assuming 270° of pointer travel). If these conditions are not met, again unmesh the sector and move the pointer counterclockwise more or less than previously. Some gauges are made so that the hairspring tension increases with increasing pressure. Therefore, it is wise to study a specific movement before disassembling the movement so that it can be reassembled in the same manner.

Hairspring tension must be sufficient to take out all the play between the gear teeth and the various linkages. Excessive hairspring tension will result in high friction, or jumpy pointer motion, or even balling of the hairspring. A little experimentation with hairspring tension is often necessary to obtain the most desirable condition. When the hairspring is properly set, go over the various pivot points and be sure there is no evidence of binding and that there is some end play.

7.7.2 Hysteresis

1. Definition. Hysteresis is the variation at any given point, between the upscale and downscale indication, after light tapping to eliminate friction in the mechanism (see Fig. 7.7). It is important to understand that hysteresis differs from friction in that when hysteresis is present in a gauge it cannot be eliminated by tapping. This condition will cause the indication to read higher on decreasing pressure than on increasing pressure, and it is a result of temporary yielding of the pressure element.

2. Correction. As noted above, hysteresis cannot be eliminated by a mechanical correction. However, the error may be split equally between increasing and decreasing pressure in order to minimize the total error. Fig. 7.8 illustrates an example of hysteresis effect in chart form.

7.7.3 Range Error

1. Definition. Range error is an error that increases in magnitude as the pointer traverses the scale from zero to maximum scale indication (see Fig. 7.9). This condition causes increasingly minus errors with increasing pressure if the bourdon is "strong," and increasingly plus errors with increasing pressure if the bourdon is "weak." Range error is caused by an incorrect distance between the center of the arbor (sector pivot point) and the centerline of the link that connects the bourdon with the sector.

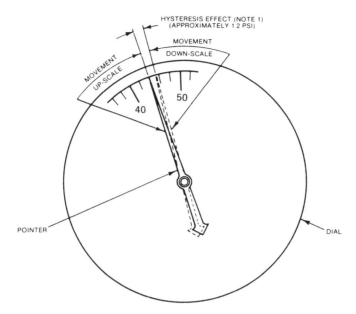

Fig. 7.7 Effect of hysteresis on indication. Deviation illustrated is hysteresis effect at 44 psi after tapping to minimize friction. The deviation is usually maximum at approximately midscale.

2. Correction. When the error in indication is increasingly plus, the pointer is said to be moving too fast, or the bourdon is too weak, in which case the link screw should be loosened and moved a little at a time in the direction B (see Fig. 7.10). If the pointer moves too slowly, the link should be moved in direction A until the indication agrees with the master. When properly set, the screw should be tightened securely. If the gauge does not have a link screw, the sector tail must be bent in the direction indicated.

7.7.4 Scale Shape Error

1. Definition. Scale shape error is a deviation in linearity that will create plus or minus errors at various points over the entire span, even if the indication at the beginning and end of the span is correct (see Fig. 7.11). The condition is a function of the linearity of the pressure element and gauge layout (see Sec. 3.4.3). If the arc of the scale on the dial is not concentric with the pinion,

Correction of Errors

Master	Error up	Error down
0	0	0
50	0	0
100	0	+0.5
150	0	-1.0
200	+0.5	-1.5
250	+1.0	+2.5
300	+0.5	+2.0
350	+0.5	+2.5
400	+1.0	+2.5
450	0	+1.5
500	+1.0	+2.0
550	0	+1.5
600	-1.0	+0.5
650	0	+1.0
700	0	+1.0
750	+0.5	+1.0
800	-0.5	0
850	-0.5	-0.5
900	+0.5	+0.5
950	0	0
1000	+0.5	0

Fig. 7.8 Hysteresis record.

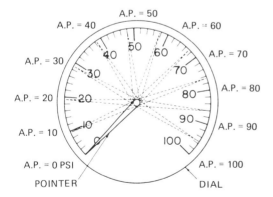

Fig. 7.9 Effect of range error on indication. This drawing illustrates the range error caused by a strong pressure element. A weak element would cause a reverse effect, i.e., at applied pressure the pointer would indicate a greater pressure than was actually applied. A.P. = applied pressure.

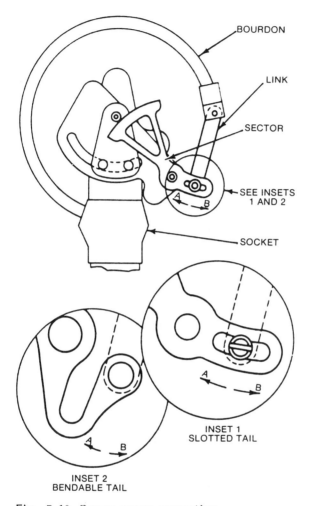

Fig. 7.10 Range error correction.

the result will be scale shape errors. Therefore, before applying the following corrections, check that the pinion is in the center of the hole in the dial and that the scale is concentric with the hole.

2. Correction. This adjustment involves rotation of the entire movement. Units that do not provide a movement rotation feature may provide for adjustment in the length of the link, which has

Correction of Errors   205

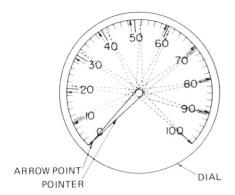

Fig. 7.11 Effect of scale shape error on indication. Dotted line pointers indicate effect of link being too long. Solid arrow points indicate effect of link being too short. Note that in both instances the magnitude of the error increases and then decreases.

essentially the same effect as rotation of the movement (see Fig. 7.12). Gauges having accuracy grades of A, B, C, and D usually do not provide any means of scale shape adjustment.

In order to make a scale shape adjustment, the complete internals, including dial and pointer, should be removed from the case as a unit. Solid front gauges may be adjusted in the case by removing the safety back, which exposes all of the adjustments. In either construction, the components and adjustments will be essentially similar.

When the errors are increasingly plus for the first portion of the scale, the complete movement should be rotated in direction A, as shown in Fig. 7.12, or the link should be shortened, as indicated at A. If the errors are increasingly slow for the first portion of the scale, the movement should be rotated in direction B, or the link lengthened, as indicated at B.

When the pointer travel is evenly divided on both sides of midspan, that is, the error is zero at 0, midscale, and fullscale, the proper angle between the link and sector tail has been established to provide the best linearity. The locking screws should now be tightened to secure this position.

7.7.5 Zero Shift Error

1. Definition. Zero shift is an error of constant value over the entire span and may be either a plus or a minus value. This condition

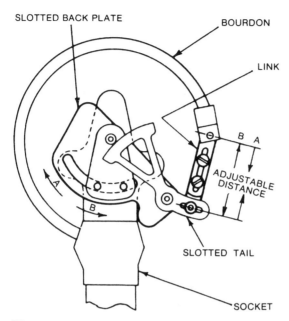

Fig. 7.12 Scale shape error correction.

may be caused by a shift of pointer position on the pinion, rotation of the dial, or a change in tip location of the bourdon.

2. Correction. Compensation for this condition is accomplished by simply repositioning the pointer or, on gauges with adjustable dials, adjusting the dial to agree with the indication of the standard. The methods of pointer adjustment were discussed in detail in Sec. 3.14.2.2. Adjustment of the dial may be accomplished by turning a knob, which usually protrudes through the case front. Some gauges are provided with dials adjustable by means of slotted mounting holes. This method necessitates removing the bezel and window, loosening the dial mounting screws, and carefully resetting the dials.

7.7.6 Determination of Required Adjustments

Examination of a set of error readings, such as would be produced by filling in the chart shown in Fig. 7.6, will often give little guidance as to the adjustments required. In such a case, plotting the

Correction of Errors

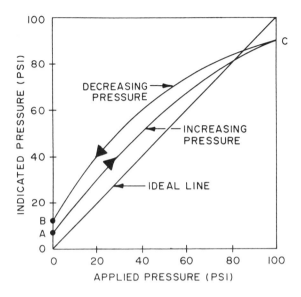

Fig. 7.13 Plot of readings—uncorrected.

actual readings will generally clarify the problem. For example, Fig. 7.13 shows a plot of the readings of a 0–100 psi gauge. (For the purpose of clarity, the actual readings rather than the errors are plotted. Where the errors are a small percentage of the actual reading, it is better to plot only the errors, since it gives a magnified view.) If indications of the gauge being tested were in perfect agreement with the true pressure, the plot would produce the straight line 0 to 100, the "ideal line." The deviation from this line represents the error. In this plot, friction, hysteresis, range, scale shape, and zero shift are all represented, although to an exaggerated degree to assist in the explanation.

The separation of points A and B would probably be due to excessive friction that could not be removed by tapping. A set in the pressure-sensing element would exhibit the same separation, but this should not occur with a properly designed element except in very high-pressure ranges. By cleaning the movement, adjusting the hairspring tension, and freeing any binding linkage, point B should coincide with point A. The displacement of point A from zero is a zero shift error and can be corrected by adjusting the pointer so as to bring point A to zero. The cause of point C being below the ideal

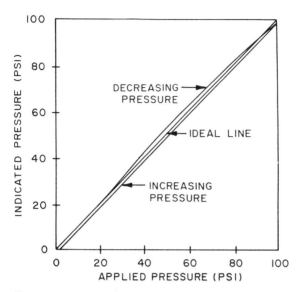

Fig. 7.14 Plot of readings—corrected.

line is a range error and is corrected as explained in Sec. 7.7.3 so as to put C at the intersection of the two 100-psi axis lines. The curvature of the lines is scale shape and can be improved as discussed in Sec. 7.7.4. By working back and forth between the various adjustments, it should be possible to obtain a plot as shown in Fig. 7.14. Note that there is still a separation between the indication of increasing and decreasing pressure. This is hysteresis and there is no adjustment which will eliminate it. By adjusting the pointer, the error can be split, which will minimize the error in either direction.

When the gauge has been successfully calibrated, all adjustment screws should be securely tightened and the gauge reassembled. No lubrication is used on the bearings. However, if the movement was cleaned in an ultrasonic bath, it may be necessary to add a film of light oil so thin as to be invisible. Pointers must be firmly seated on the pinions and spaced sufficiently away from the dial and window to insure there will be no interference. If the pointer is replaced it may be necessary to taper ream the pointer bushing to get it to seat far enough down on the pinion to clear the window.

Unless specifically trained, one should not attempt to repair damage to the pressure-containing envelope. Replacement is the best solution. The method of replacing windows, bezels, etc. will generally be obvious and need not be discussed here.

7.8 MAINTENANCE OF DIAPHRAGM SEALS

If the maintenance shop will be required to assemble or rework gauges having diaphragm seals, it is necessary to provide a special filling apparatus. Basically, the procedure consists of evacuating and filling the pressure gauge, filling the chemical seal, joining the two, and calibrating the assembly.

7.8.1 Equipment

A suggested arrangement is shown in Fig. 7.15. The filling chamber is made in the form of a sight glass. The capacity of the sight glass depends on the volume and quantity of the parts to be filled, but, as a guide, 300 to 500 cc is sufficient. A fill fluid reservoir, having a capacity of about 1 liter, and a surge tank are connected to the sight glass as shown. The surge tank, which is made from a pipe nipple 6 in. long and 2 in. in diameter, prevents the fill fluid from being drawn into the vacuum pump. When a vacuum is initially applied to the fill chamber, considerable foaming of the fill fluid takes place due to the escape of air dissolved in the fluid and to air from the gauges being filled bubbling up through the fluid. Many fill fluids will vaporize (boil) at room temperature when subjected to very low pressure, and it will be impossible to obtain a pressure less than that corresponding to the boiling point. Therefore, the vacuum pump used in the system only needs to be capable of maintaining a pressure of about 0.2 in. Hg absolute. Valve A of Fig. 7.15 should be a needle valve (see Fig. 5.11). This type will permit a controlled application and release of the vacuum to the system, thus minimizing foaming of the fill fluid and preventing the gauge mechanism from being driven rapidly against the minimum stop. Valve B is a three-way, two-position valve arranged so as to either connect the system to the vacuum source or vent it to atmosphere. Valve C should have a maximum opening such as provided by a stopcock valve (Fig. 5.12) or a ball valve. The piping from the top of the sight glass to the vacuum pump may be tubing 1/4-in. in diameter. A manifold for mounting the gauges to be filled is connected to the bottom of the

210   Chap. 7 Gauge Maintenance and Calibration

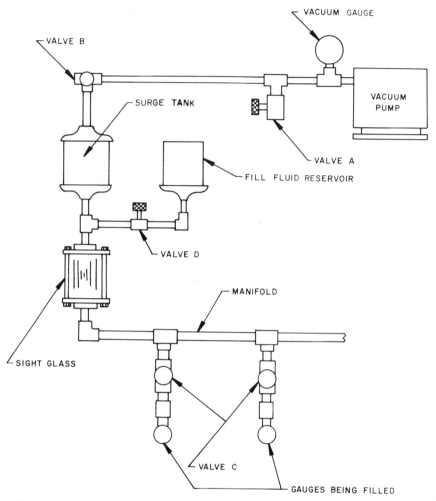

Fig. 7.15 Diaphragm seal filling apparatus.

sight glass using tubing having at least a 1/2-in. bore. The arrangement, size, and number of filling ports on the manifold can be varied to suit the needs of the maintenance shop. The gauge should be mounted to the manifold in an inverted position. At the start of the evacuation process, the air within the gauge to

Maintenance of Diaphragm Seals

be filled must be drawn into the manifold and through the fill fluid. Therefore, the vertical distance between the surface of the fill fluid in the sight glass and the gauge port should be kept as short as possible, and the piping and valves should be arranged so that air bubbles working their way through the fluid will not be restricted or trapped. Of course, it is essential that all joints be leak-tight. The system should be checked periodically by evacuating it, shutting off the source of vacuum, and noting any change in indication of the vacuum gauge over a period of 15 to 30 min.

7.8.2 Filling the Gauge

The procedure for filling the gauge is as follows:

1. Remove any checks or dampers from the gauge port. Attach the gauge to be filled to the manifold. Close valves C.
2. Fill the reservoir with fluid. Position valve B to vent the system to atmosphere, and then open valve D to allow the fluid to flow into and fill the sight glass about half full. Close valve D and position valve B to connect the system to the vacuum source.
3. With valve A open so as to vent the line to atmosphere, start up the vacuum pump. Open valves C and slowly close valve A. As vacuum is applied to the system, the fill fluid in the sight glass will rise rapidly and considerable bubbling and foaming will take place as previously noted.
   At first it may be necessary to leave valve A partially open to keep the fluid from entering the surge tank. Gradually close valve A until it is fully closed. When the foaming and bubbling stop (usually after a matter of several minutes), release the vacuum by opening valve A to atmosphere.
4. The level of fill fluid in the sight glass will drop as the fluid flows into the evacuated pressure envelope of the gauge. When the level stabilizes, apply vacuum again by closing valve A. The level of the fluid in the sight glass will remain essentially the same if a good fill was attained. If air remains in the system, it will expand under the reduced pressure and cause the fluid level in the sight glass to rise. If this occurs, continue to alternately evacuate and fill the system until the fluid level remains at a fixed level regardless of whether valve A is open or closed.
5. When this is achieved, position valve B so as to shut off the vacuum source and vent the system to atmospheric pressure. Close valve C, remove the gauge, and store it with the pressure connection in an upright position. Replace any checks or dampers removed from the gauge prior to filling.

7.8.3 Filling the Diaphragm Seal

1. Seals with metallic diaphragms and without a fill/bleed screw. Fig. 5.1 illustrates a seal having a fill/bleed screw. If the seal does not have one, it must be filled through the gauge port. A convenient way to do this is by means of an oil can. Clamp the seal in a vise and work out any bubbles by gently flexing the diaphragm from the process side using a finger or the eraser end of a pencil. Fill the seal to the top of the port.

An alternative method of removing air is to rapidly rotate the filled seal. The fill fluid will move to the periphery, thus forcing any air bubbles to the center, where they will escape through the gauge port.

Seals that are to be exposed to high temperature can be baked to drive out any air dissolved in the fluid. Seals to be used for vacuum applications should be evacuated after filling to remove any entrapped or dissolved air. Evacuation should be done in a chamber so that the vacuum will be applied to both sides of the diaphragm simultaneously to avoid overstressing it.

2. Seals with metallic diaphragms and with a fill/bleed screw. If the seal is equipped with a fill/bleed screw, it may be filled as described above (Sec. 7.8.3.1) or it may be attached to the gauge before filling and the assembly of the gauge and seal filled as a unit. To accomplish this, the filling apparatus shown in Fig. 7.15 is used, except that a short length of tubing (preferably clear polyethylene) is attached to a station on the manifold. The other end of the tube is fitted with a connection that will mate with the thread of the fill/bleed port. The evacuation and filling procedure is as described in Sec. 7.8.2.

If the seal does not have a contoured surface on the inner face of the upper housing (Sec. 5.2.6.4), the diaphragm could be damaged during evacuation due to the atmospheric pressure on the process side. In this case it would be best to fill the seal before joining it to the gauge. Alternatively, the vacuum can be applied to both sides of the diaphragm.

3. Nonmetallic seals. Seals utilizing nonmetallic diaphragms such as Teflon or Viton are filled through the gauge port in the same manner as those using metallic diaphragms. Teflon diaphragms are translucent, and any air bubbles remaining in the seal will be visible if the seal is turned upside down and the diaphragm viewed from the process side. Viton diaphragms are soft and will sag under the weight of the filling fluid. Applying an air pressure of a few inches of water to the process side will support the diaphragm during the filling process and prevent adding an excess of fluid. By varying

Maintenance of Diaphragm Seals

the pressure the diaphragm can be gently flexed to assist in air removal.

7.8.4 Joining the Gauge and Seal

The pressure port of the gauge must be completely full. Add fluid if necessary. Apply a thread sealant to the pressure gauge connection. The fill fluid that is displaced when the pressure gauge is threaded into the seal will deflect the diaphragm outward, which is undesirable. To prevent this the diaphragm must be deflected toward the upper housing so as to expel an amount of fluid approximately equal to the volume of the pressure gauge connection that will enter the seal. If the inner face of the upper housing is contoured to match the diaphragm convolutions, apply an air pressure between 5 and 10 psi to the process side of the diaphragm to expel the fluid. If the housing is not contoured, then a maximum of 3 psi should be applied. Higher pressure will distort and possibly destroy the diaphragm. Thread the gauge into the seal while maintaining the air pressure on the diaphragm. Entry of the connection should push the diaphragm away from the upper housing so as to bring it to its unstressed position when the connection is tightly wrenched in place.

It will be appreciated that bringing the diaphragm to its neutral or unstressed position using the above method is difficult and uncertain due to variations in taper pipe threads. One seal manufacturer has designed a special fitting for use with their contoured seals for controlling and obtaining the proper fill (see Fig. 7.16). In use, the union nipple is attached to the seal. A union body is attached to the gauge. Both halves of the union are filled, pressure is applied to the process side of the diaphragm, and the union body is inserted in the union nipple. The two O-ring seals on the union body produce a controlled displacement to return the diaphragm to its neutral or unstressed position.

If the seal is equipped with a fill/bleed screw, the joining process should be done so that the diaphragm is deflected slightly toward the process side. After the gauge is installed on the seal, the fill/bleed screw may be opened to expel excess fluid and allow the diaphragm to return to its unstressed position, after which the screw is resealed.

When joining the gauge to seals using elastomeric diaphragms, the same procedure is used as for seals with metallic diaphragms. If the seal is to be used on a positive-pressure application, the pressure applied on the process side during filling should be controlled so as to support the diaphragm without deflecting it against the upper

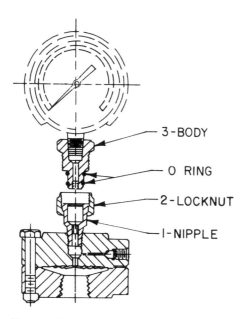

Fig. 7.16 Union connection — diaphragm seal. (Courtesy of Mansfield & Green Division, AMETEK, Inc.)

housing. If the seal is to be used for vacuum applications, pressing the elastomeric seal against the upper housing while assembling the gauge is beneficial because the diaphragm will be capable of moving a greater distance to transmit the vacuum to the gauge.

It is probable that some loss of accuracy will occur due to the diaphragm seal, particularly if the diaphragm is metallic and the pressure is low. Therefore, it is necessary to calibrate the assembled gauge and seal after filling to determine if the accuracy is acceptable for the application. If it is not, zero and range adjustments may be made to the gauge to reduce the errors.

The ratio of volume change to an incremental pressure change is much greater for diaphragm and bellows pressure elements than for bourdons. The deflection of the diaphragm in the seal must provide the volume change, and therefore it is not always possible to use a seal in conjunction with bellows or diaphragm pressure elements. Such applications should be discussed with the seal manufacturer. Additional information on filling equipment and filling techniques can also be obtained from the same source.

8
Pressure-Actuated Thermometers

8.1 INTRODUCTION

Pressure gauges may be adapted to indicate temperature by measuring the pressure change of a contained fluid (liquid or gas) with changes in temperature. Three basic systems are commonly used:

Gas-filled
Liquid—solid-filled
Liquid—vapor-actuated

8.2 GAS-FILLED THERMOMETERS

8.2.1 Theory

Gas-filled thermometers utilize a bourdon-actuated pressure gauge as the indicating head. A bulb is connected to the gauge using an appropriate length of small-bore capillary, and the entire system is charged with an inert gas, usually nitrogen. The bulb is subjected to the temperature to be measured, and the expansion of the gas with increasing temperature will raise the pressure within the system; conversely, the contraction of the gas with decreasing temperature will lower the pressure in the system.

216   Chap. 8 Pressure-Actuated Thermometers

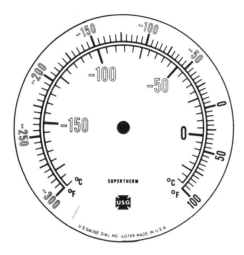

Fig. 8.1 Typical gas-filled thermometer dial. (Courtesy of AMETEK, Inc.)

8.2.2 Volume of Bulb

If the volume of the bulb is large compared to the volume of the capillary and bourdon, the pressure change with temperature will approximate the relationship discussed in Sec. 4.2 and be very nearly linear, resulting in a scale having essentially evenly spaced graduations for indicating the temperature. Fig. 8.1 shows a typical scale layout for a gas-filled thermometer. Note that the scale is slightly nonlinear, the interval from 0°C to 25°C being smaller than from -150°C to -175°C. This is mainly the result of a small ratio of bulb volume to bourdon plus capillary volume. The linearity could be improved by using a larger bulb if the application permits.

8.2.3 Ratio of Bulb Volume to System Volume

It is also desirable to keep a high volumetric ratio between the bulb and the indicating head plus capillary in order to minimize the effect of ambient temperature changes at the head and along the capillary. If, for example, the ratio is 30:1, then a 30° increase in temperature at the head and along the capillary will increase the pressure in the

Gas-Filled Thermometers 217

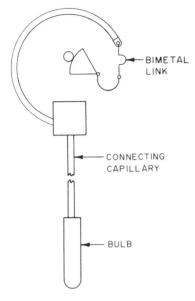

Fig. 8.2 Head temperature compensation — gas-filled thermometer.

system approximately the same amount as a 1° increase in the temperature at the bulb. If the magnitude of the effect of ambient temperature change at the head and along the capillary is not acceptable, then some means of compensation must be used.

8.2.4 Temperature Compensation

If the effect of temperature changes only at the head is to be compensated, it is possible to add to the bourdon a bimetal link that will impart motion in the direction opposite to that imparted by the expansion or contraction of the gas in the bourdon due to ambient temperature changes (see Fig. 8.2).

If the effect of the capillary must be compensated, it is necessary to employ a second filled system comprising a bourdon and capillary but without a bulb. The second system is arranged to move in a direction opposite to the primary system and thereby subtract any motion that results from either head or capillary temperature changes (see Fig. 8.3).

218   Chap. 8 Pressure-Actuated Thermometers

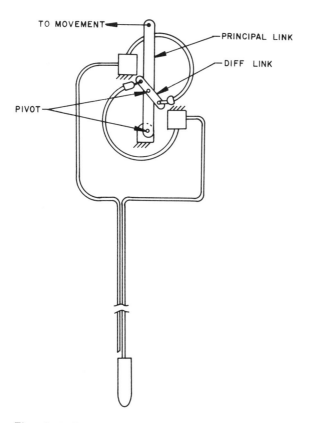

Fig. 8.3 Head and capillary temperature compensation — gas-filled thermometer.

8.2.5 Gas-Filled Thermometer Considerations

For the reasons discussed above, bourdons for gas-filled thermometers are made with a small minor axis and the bulbs are very large, on the order of 10-in. long by 3/4-in. in diameter. The pressure change with temperature is independent of the bulb size, and various temperature spans are obtained by varying the initial charging pressure and/or the pressure range of the bourdon.

The advantages of gas-filled thermometers include an essentially linear scale, no cross-ambient effect (see Sec. 8.4.4), ability to be used at very low temperatures, good inherent accuracy, reasonable cost, and a nontoxic filling medium. A disadvantage is that it is difficult to make a gas-filled thermometer with a short temperature span. The reason is that the pressure change for a small temperature change is not a sufficient portion of the total pressure to provide adequate motion of the bourdon. For example, a 0–300°F thermometer may have a fill pressure of 140 psi at 0°F and 241 psi at 300°F. Therefore, the bourdon must develop sufficient motion to move the indicating pointer from 0°F to 300°F for a pressure change of 101 psi. Note, however, that the total pressure at 300°F is 241 psi or 2.4 times the pressure change. If the bourdon must develop a motion of, say, 0.1-in. to move the pointer, then the total motion of the bourdon must be 0.24 in., with the usable portion being only that which is between 0.14 and 0.24 in. or 140 and 241 psi. Such large motion imposes high stress levels in the bourdon. Shortening the span to 100–300°F would of course make the problem even more severe.

8.3 LIQUID-FILLED THERMOMETERS

8.3.1 Theory

A system having the bourdon, capillary, and bulb solidly filled with a liquid may also be used as a temperature indicator. Strictly speaking, however, these are not pressure-actuated devices. Rather, they are volume-actuated, because they depend on the change in volume of the fill-fluid with temperature to actuate the bourdon. As the volume of the fluid in the bulb increases the walls of the bourdon are pushed outward (in the same manner as if the bourdon were pressurized), and therefore the tip of the bourdon moves. The liquid may be considered incompressible, in which case the bourdon must elastically deform until it can accommodate the new volume of the liquid irrespective of the pressure generated. The larger the bulb, the more the volume change will be for a given temperature change, and therefore the greater travel the bourdon will have for that temperature change. The temperature span of the system is therefore a function of the bulb volume, the change in bourdon volume incurred as the bourdon moves through its travel, and the temperature coefficient of expansion of the fill fluid.

Fig. 8.4 shows a typical dial for a solid-filled system. Note that the increments are linear over the entire range.

8.3.2 Volume of Bulb

Mercury is often used as the fill fluid in solid-filled systems because it has a wide usable temperature range of -40°F to 1200°F. It has a

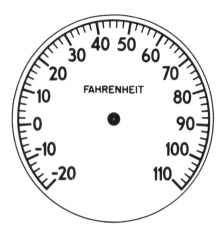

Fig. 8.4 Typical liquid-filled thermometer dial. (Courtesy of AMETEK, Inc.)

low coefficient of expansion (compared to other liquids) which requires larger bulbs but permits a longer length of capillary to be used before it becomes necessary to compensate the capillary. Various organic liquids, such as toluene, may be used in place of mercury. Toluene has about six times the temperature coefficient of expansion of mercury, and therefore requires a much smaller bulb for a given temperature span. A typical bulb size for a mercury-filled thermometer having a span of 0–100°F is 3 1/2-in. long by 1/2-in. in diameter. For a toluene-filled thermometer in the same span, the bulb size would be 2-in. long by 1/4-in. in diameter.

8.3.3 Temperature Compensation

Compensation for ambient temperature changes at the indicating head or along the capillary may be accomplished in a similar manner as with gas-filled thermometers. Because of the low coefficient of expansion of the mercury, it is also possible to compensate the capillary by inserting into the bore of the capillary a wire made from a material having a very low coefficient of expansion. By properly choosing the size of the wire in relation to the capillary, the net effect of the expansion of the capillary and wire can be made to equal that of the mercury.

Vapor-Pressure Thermometers

8.3.4 Mercury Fill

Mercury-filled thermometers are not permitted in certain applications, for example, food processing, where the contamination of the process medium that would occur if the filled system ruptured would create a hazard. For food processing applications, a compatible filling medium such as mineral oil or a glycerin and water mixture is often used. Because of the toxicity of mercury vapor, the use of mercury for thermometry is declining. Mercury-filled thermometers require steel or stainless steel for parts wetted by the mercury.

8.3.5 Solid Fill Considerations

Solid filled systems are offered in temperature spans as small as 100°F. When the system is filled, sufficient pressure is applied to the filling medium to be certain that when the bulb is subjected to the coldest temperature within the range of the thermometer, and the volume of the liquid is at its minimum, there will still be a positive pressure in the system. Further, the fill pressure must be higher than the vapor pressure for the filling medium at the maximum temperature of the span.

8.4 VAPOR-PRESSURE THERMOMETERS

8.4.1 Theory

Vapor-pressure-actuated thermometers (sometimes called vapor tension thermometers) use a volatile liquid such as ethyl ether for the fill and measure the vapor pressure of the liquid corresponding to the temperature at the bulb. Vapor-pressure-actuated thermometers are the most widely used of the filled-system type and so will be discussed here in greater detail than the gas- or liquid-filled types.

8.4.2 The Nature of Vapor Pressure

A short discussion of what is meant by vapor pressure will be helpful. Matter may exist as a gas, a liquid, or a solid, depending on the temperature and pressure to which it is exposed. Water, for example, is normally considered a liquid because it exists as a liquid in most of the atmospheric temperature and pressure conditions in which we encounter it. Yet we know that if we subject it to a temperature of less than 32°F it will become a solid, and if we heat it above 212°F it will boil and produce water vapor, which is a gas. Examining the boiling point further, we know that it will vary with atmospheric pressure. At sea level the boiling point is 212°F, but at an altitude of 8000 ft water will boil at a temperature of about 197°F. On the other hand, if water is boiled in a pressure cooker it will reach a temperature of 226°F if the cooker pressure is maintained at 5 psi. It can be said, therefore, that when the temperature of a liquid is increased until it boils it will have created a pressure, called vapor pressure, which is equal to the surrounding pressure. Conversely, the surrounding pressure is determined by the vapor pressure of the boiling liquid. If, as with an open pot of boiling water, the surrounding pressure does not exceed 14.7 psia, then the temperature of the water will not exceed 212°F. However, if the water vapor is trapped, as in a pressure cooker, then the pressure above the water will increase as vapor is driven off, and the boiling point of the water will also increase. The equilibrium temperature of the boiling water will be determined by either the pressure at which the cooker is vented or the rate at which heat is applied to the cooker as opposed to the rate at which heat is lost from the cooker. For example, if the source of heat is low, then at some point the units of heat entering the cooker will be equalled by the units of heat being removed from the cooker, and the water temperature and pressure will stabilize. Increasing the heat source will cause the cooker to stabilize at a higher temperature and pressure, and a large increase could create sufficient pressure to rupture the cooker if the venting means became inoperative. The bulb of a vapor-actuated thermometer acts like a small sealed pressure cooker. When it is immersed in a process fluid that is above ambient temperature, the filling liquid within the bulb will be raised to the temperature of the process fluid and thereby create a vapor pressure that is related to the temperature of the process. If the process is below ambient temperature, then the filling liquid will be cooled to the temperature of the process fluid, again creating a pressure that is related to the temperature of the process.

8.4.3 Pressure Versus Temperature

The vapor pressure/temperature relationship varies depending on the liquid used, and it is necessary to select the liquid to yield a usable pressure change over the desired temperature span. Fig. 8.5 shows the vapor pressure/temperature relationship for two liquids, ethyl ether and propane. In general, ethyl ether is used to measure temperatures between 100 and 350°F, while propane is useful at temperatures between -40 and 200°F.

Reference to Fig. 8.5 shows that the pressure/temperature relationship is very nonlinear. For example, if the thermometer is

Vapor-Pressure Thermometers

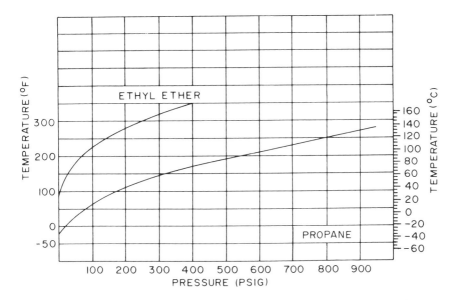

Fig. 8.5 Vapor pressure versus temperature. (Courtesy of AMETEK, Inc.)

filled with ethyl ether, the pressure at 100°F will be 5.0 psi, and for 150°F it will be 27 psi. This is a change of 22 psi for a 50°F temperature change. The pressure at 200°F will be 70 psi, and at 250°F it will be 139 psi. This is a 69-psi change in pressure for a 50°F change in temperature. Since the bourdon responds linearly to pressure, the pointer travel from 200 to 250°F will be 70/22, or 3.2 times the pointer travel from 100 to 150°F. The scale of a typical vapor pressure thermometer is shown in Fig. 8.6. The compression of the scale at the low-temperature end of the range is characteristic. For this reason, the span of vapor pressure thermometers should be selected so that the point of interest is at least at midscale in order to obtain good readability.

8.4.4 Cross-Ambient Effect

Vapor pressure thermometers have another disadvantage called the "cross-ambient effect," which makes a vapor-pressure-actuated thermometer unreliable for the measurement of temperatures near

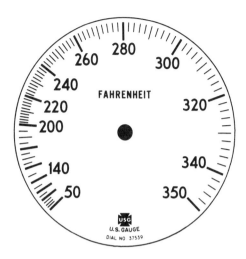

Fig. 8.6 Typical vapor pressure thermometer dial. (Courtesy of AMETEK, Inc.)

the ambient temperature at the indicating head. In order to explain the cross-ambient effect it is necessary to understand what occurs inside of the filled system as the temperature at the bulb varies. Consider the bulb of a vapor pressure thermometer that has just been immersed in a water bath at 200°F while the head is at 70°F. For the purpose of the explanation we can assume the sequence of events occurs in steps, although in actuality it is a continuous process. The filling medium in the bulb (for this example let us use propane) will boil and drive off propane vapor. The pressure will rise in order to establish the vapor pressure of propane at 200°F (approximately 559 psi). Vapor at 559 psi flows out of the bulb to the capillary and indicating head. The temperature of the capillary and head is 70°F, and the vapor pressure of propane corresponding to 70°F is only 111 psi. Therefore, the vapor will condense to a liquid, lowering the pressure in the system. Since the pressure in the bulb is now less than 559 psi, additional vapor will be boiled off and again condense in the capillary and head. The process of vaporizing the propane in the bulb and condensing it in the cooler part of the system will continue until the entire capillary and head are filled with liquid propane, at which point an equilibrium condition will be established, the pressure in the bulb will stabilize at 559 psi, and the pointer will indicate 200°F.

Vapor-Pressure Thermometers

Now suppose the water bath is allowed to cool. As the temperature of the propane decreases, the pressure in the bulb decreases and the pointer will follow the drop in bath temperature. When the bath reaches a temperature below the head temperature, the bulb becomes the coolest part of the system and the vapor in the bulb will condense, lowering the pressure in the system to a value less than 111 psi (the vapor pressure of propane at 70°F). The liquid in the head and capillary will therefore begin to boil in an attempt to maintain the 111 psi. The higher pressure vapor will flow into the bulb, where it condenses. Again, the process continues until all of the liquid in the head and capillary is vaporized, at which time all the liquid in the system is contained in the bulb. The transfer of filling liquid from the bulb to the capillary and head that occurs when the bulb temperature increases from below ambient to above ambient, and the opposite condition, where the filling liquid transfers from the capillary and head to the bulb as the temperature of the bulb decreases from above ambient to below ambient, require time to take place. While the transfer process is occurring, the indicating pointer will hover around the ambient temperature graduation, moving above or below it in a random manner, regardless of the actual bulb temperature. If the bulb temperature remains in the vicinity of ambient, moving several degrees above and below it, the pointer will continue to give an erratic indication. The cross-ambient effect is the lag in response time and the erratic pointer motion at ambient temperature that occur due to the transfer of liquid.

8.4.5 Volume of Bulb

It will be apparent from the explanation given in the previous section that there must always be some liquid and some vapor in the bulb in order to be certain that the temperature at the bulb controls the pressure in the system. Because of this, the volume of liquid fill in the system must be equal to the sum of the internal volume of the indicating head, the capillary, and a portion of the bulb if the thermometer is to indicate temperatures above ambient. Further, the volume of the bulb must be capable of containing all of the fill fluid if the thermometer must indicate temperatures below ambient.

The volumetric expansion of the liquid must be considered when charging the system. If, for example, the system is almost completely filled with liquid when it is charged at room temperature, the expansion of the liquid at higher temperatures may completely fill the system, and it will no longer act as a vapor pressure thermometer but instead as a solid-filled liquid thermometer, and gross errors of indication will result.

To accommodate all these constraints it is customary to size the bulb so that it has a volume approximating at least twice the volume of the head and capillary. In this way the volume of liquid charged into the system need not be precisely controlled.

8.4.6 Temperature Compensation

While vapor pressure thermometers have the disadvantages of a nonlinear scale and the cross-ambient effect, they are widely used because they have the distinct advantages of not requiring either head or capillary compensation and of being inexpensive to manufacture compared to gas- and solid-filled systems. The explanation given in Secs. 8.4.4 and 8.4.5 will show why compensation is not required. Assume that the system is stabilized with the bulb at 200°F and the head and capillary at 70°F, in which case the head and capillary are full of liquid. If the temperature at the head and/or capillary increases to a temperature not exceeding the bulb temperature, then the liquid will expand but not boil. The pressure that results from this expansion is relieved by the transfer of excess liquid into the bulb, reducing the vapor space in the bulb but not increasing the system pressure. As another example, suppose that the capillary passes through an area where the temperature is higher than the bulb. The liquid at that point will boil, creating a vapor pressure higher than that in the bulb. The increased pressure will simply push liquid into the bulb until the pressure in the capillary comes into equilibrium with the pressure in the bulb. Of course the opposite is true: that is, if the liquid in the head contracts due to a lower ambient temperature, the pressure in the system will decrease. The liquid in the bulb will then be at a pressure less than that corresponding to the temperature of the bulb and therefore additional vapor will be boiled off to restore the pressure. Note again that it is necessary to size the bulb and the volume of filling liquid so that there will always be an interface of liquid and vapor in the bulb under all possible conditions of bulb, head, and capillary temperature. If there is too little liquid in the system, the bulb will be depleted of liquid (if it is at a higher temperature than the capillary and head) and the liquid vapor interface may occur somewhere in the capillary, in which case the thermometer will indicate the temperature at the point of the interface. If, however, there is too much liquid in the system, the bulb will not be able to contain all the liquid (if it is at a lower temperature than the head and capillary) and the liquid vapor interface will occur somewhere in the capillary, and the temperature at this point will be the temperature indicated.

Effect of Bulb Position 227

8.5 EFFECT OF SHIPPING TEMPERATURE

All pressure-actuated thermometers must be capable of withstanding the highest temperature that may be encountered during shipping or storage, which may be as high as 160°F. If the maximum temperature of the span is higher than 160°F, there will be no problem with gas-filled thermometers. There may be a problem with vapor pressure and solid-filled thermometers in that the liquid in the entire system will be at the high temperature rather than just that in the bulb, and the system must be capable of accommodating this maximum volume without overstressing the bourdon. Thermometers having the maximum temperature of the span under 160°F may be severely overpressured by high shipping and storage temperature and should be protected against such damage. By carefully controlling the amount of filling fluid charged in a vapor pressure system, it is possible to completely vaporize all the liquid in the system at some temperature slightly higher than the maximum indicated temperature. At this point, the rate of pressure change with temperature will follow the gas law and be substantially decreased so that the thermometer may be shipped without damage resulting from excessive temperature.

8.6 EFFECT OF AMBIENT PRESSURE

The indication of both gas-filled thermometers and vapor pressure thermometers is affected by ambient pressure changes. The magnitude of the effect will of course depend on the ratio of pressure change to temperature change for the particular thermometer being used. Normal barometric extremes can usually be ignored, but if the thermometer is used at an altitude substantially above sea level the error may be significant. If there is a possibility of significant errors due to ambient pressure change, the user should have the manufacturer determine the actual error expected. Solid-filled liquid systems, as previously stated, are volumetric devices and therefore not subject to errors due to ambient pressure changes.

8.7 EFFECT OF BULB POSITION

If the bulb of a vapor pressure thermometer is positioned substantially above or below the pressure-sensing element, it may be necessary to correct for the pressure effect of a liquid head. For example, if the bulb is above ambient temperature, the pressure element and capillary will be full of liquid. If the pressure element is positioned above the bulb, the pressure effect of the liquid column will subtract from the pressure in the bulb, causing the temperature indication to be low. If the pressure element is positioned below the bulb, the liquid column will add to the pressure in the bulb and cause the temperature indication to be high. The magnitude of the error will depend on the vapor pressure at which the thermometer is operating and the vertical distance between the bulb and the pressure element. The actual effect can be determined at installation by raising and lowering the position of the gauge and noting any change in indication. Because the pressure effect of the head is a constant, it can be compensated for by adjusting the pointer. If the bulb is below ambient temperature, the pressure element and capillary contain only vapor and there is no head effect.

Vapor pressure thermometers that have cross-ambient spans cannot be compensated for liquid head effect. This is because the filling liquid will be in the pressure element and capillary if the bulb is above ambient, thus requiring compensation, or in the bulb if the bulb is below ambient, thus requiring no compensation.

The liquid head effect is usually negligible in thermometers having solid-filled liquid thermal systems. The length of capillary connecting the head and the bulb is rarely sufficient to create a head pressure larger than the system charging pressure and the compressibility of the fill fluid and the compliance of the bulb and capillary can be assumed to be nil. In mercury-filled systems where the head exceeds about 30 ft, it may be necessary to make some correction and the supplier should be consulted.

9
Pressure Transducers

9.1 INTRODUCTION

9.1.1 Definition

A pressure transducer is a device that will convert an applied pressure to some other measurable form of energy. Under this definition, a bourdon pressure gauge might be considered a transducer since it converts a pressure change to a measurable motion. However, as used in the industry, the term "pressure transducer" usually refers to a device that converts the pressure change to a change in some electrical quantity (voltage, current, resistance, capacitance).

Other allied terms are "pressure sensor," "pressure transmitter," and "pressure sender." The distinction among these terms has become blurred over the years, but in general, "pressure sensor" refers to the basic device (i.e., strain gauges mounted on a diaphragm plate), whereas "transducer" refers to the completed device, including electronic circuitry for temperature compensation, power supply conditioning, and output signal conditioning. "Pressure transmitter" is most often used in the process control industry and "pressure sender" in the automotive industry. In this chapter we will simply refer to all devices of this type as "transducers."

9.1.2 Scope

Pressure transducers constitute a very broad subject, and a wide variety of transducers and transduction principles is in use. We

9.1.3 Transfer Function

The relationship between the input pressure and the output signal of a transducer is called the transfer function. In most cases the transfer function is linear. That is, a 10% of span change in the input pressure will produce a 10% of span change in the output signal. However, depending on the application, nonlinear output signals may be acceptable, or even preferred, so long as the nonlinearity is known and repeatable.

9.2 RHEOSTATS

9.2.1 Basic Circuit

A very simple pressure transducer may be made by driving the wiper arm of a rheostat with a pressure element (bourdon, diaphragm, or bellows) and, using an ammeter, measuring the change in current flow resulting from the resistance change (see Fig. 9.1). The ammeter may be calibrated in units of pressure. Note that the current flowing through the ammeter is affected not only by the change in resistance of the rheostat, but also by the applied voltage. Therefore, in order to insure the accuracy of indication, the voltage supply must be held constant. Also, changes in resistance due to temperature variations will affect the current flowing in the system.

9.2.2 Automotive Applications

Such a simplified system is rarely used, but a modified version is commonly used on many automotive and off-highway vehicles to remotely indicate engine oil pressure, hydraulic pressure, and transmission pressure (see Fig. 9.2). This version uses a meter movement that has at least three separate coils all wound on the same bobbin. A disc-shaped permanent magnet is journaled within the core of the bobbin so that it may rotate freely and align its magnetic axis with the resultant field produced by the three wound coils. A sender, located at the point of pressure measurement, has a pressure-sensitive element arranged to move a wiper over a resistance element.

The system operates as follows:

1. If the wiper is in position 1, then there is a minimum resistance in the total circuit and maximum current will flow through

Rheostats

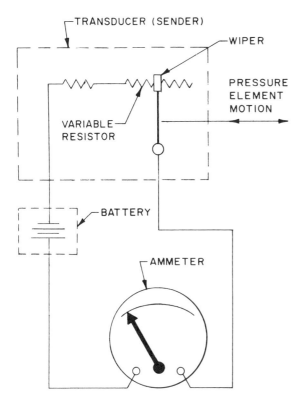

Fig. 9.1 Rheostat as a transducer.

coil X. However, only a small portion of the total current will flow through Y and Z, because there is a parallel path to ground through the sender, which is a lower resistance path. A summation of the magnetic vectors would therefore be as shown in A of Fig. 9.2, and the rotating magnet within the core of the bobbin will rotate to align its poles with the resultant vector. A pointer attached to the same shaft that carries the magnet will indicate a starting position, for example, 0 psi.

2. Pressure applied to the sender will move the wiper to position 2, substantially increasing the resistance to ground through the sender and increasing the resistance of the total circuit so that less current flows through the coil X. However, since the resistance of the path to ground through the sender has increased, more current

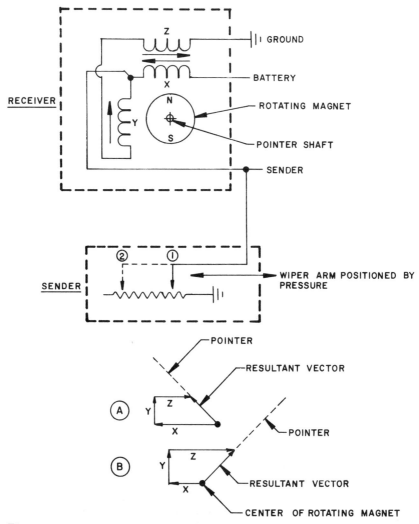

Fig. 9.2 Automotive applications. Arrows represent direction of magnetic field produced by the coil.

will flow through coils Y and Z. A summation of the magnetic vectors would be as shown in B of Fig. 9.2 and the rotating magnet will now align itself with the new vector, carrying the pointer through about 100° of arc and indicating a maximum pressure, for example, 100 psi.

With this arrangement, normal variations in the battery voltage will not affect the meter indication because all vectors are increased or decreased proportionally, resulting in the same angular position of the resultant vector, although at a smaller or larger magnitude. Such a meter is said to be insensitive to voltage supply variations. Also, note that if the sender and one end of the meter coils are grounded to the vehicle frame (the usual practice), then only one wire is required between the sender and the meter.

Because of their widespread use in automotive and off-highway applications, the meters and senders (transducers) are readily available and inexpensive. However, the system accuracy is on the order of ±5% and as much as 2 W of power may be required. Because of the high current drawn (up to 160 mA in a 12 V system), the separation between the meter and the sender is limited to about 25 ft. The meter offers a rugged construction able to withstand the vibration, shock, and temperature variations associated with automotive and off-highway use.

Obviously, these variable resistance automotive-type pressure transducers may be used with other types of readouts or for the control of some process, provided their rather poor accuracy is acceptable.

9.3 POTENTIOMETERS

9.3.1 Basic Circuit

Potentiometric transducers are similar to those described above, in that a wiper arm traverses a resistive element (see Fig. 9.3). However, a voltage is applied across the resistive element and a variable voltage is developed at the wiper. The ratio of the variable voltage to the applied voltage remains constant even if there is a wide variation in the applied voltage or a change in the resistance of the resistive element due to temperature changes. To take advantage of this feature, ratiometric voltmeters are available to read out potentiometers.

The value of the resistance used in the transducer is usually kept high (for example, 1000 Ω) in order to reduce the power consumption. The indicating device must therefore have a high resistance in order to prevent loading the potentiometer (that is, draw too much current across the resistive element), which would result in a nonlinear voltage output. A typical indicator is a high-impedance voltmeter (analog or digital) that is calibrated in terms of pressure.

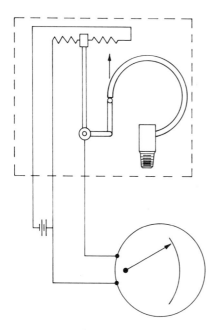

Fig. 9.3 Potentiometer as a transducer.

9.3.2 Resolution

Wipers operating over wire-wound elements produce a step change in resistance as the contact point of the wiper moves from one turn of the resistive winding to the adjacent turn; that is, the smallest change in resistance that can be obtained is equal to the resistance of one turn of the winding. If the total number of turns traversed by the wiper for the pressure span is 200, then the smallest change in pressure that can be reliably measured will be one part in 200, or 1/2%. This is called the "resolution" of the transducer; i.e., the smallest pressure change that can be resolved is 1/2% of the total pressure span. Increasing the wiper travel or decreasing the wire size used for the resistive winding will improve the resolution; that is, enable the device to resolve a smaller percentage of the total span. For example, if the travel were doubled or the diameter of the resistance wire were halved, then the resolution in the above example would be 1/4%. However, the use of more turns and

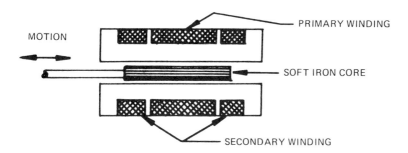

Fig. 9.4 Linear variable differential transformer (LVDT).

smaller-diameter wire increases the resistance of the winding and reduces the amount of current that can be drawn by the device. Where it is desired to drive indicators or equipment that require higher current (e.g., valves, servo motors, etc.), it is necessary to provide an amplifier or relay.

9.3.3 Conductive Plastic Elements

Wire-wound resistors and wipers are subject to mechanical wear, especially if used in an environment of pulsing pressure or high vibration. Thick-film resistive elements, or elements made of an electrically conductive plastic, provide better wear characteristics. Since the wiper, when moving over such a resistive element, does not move from turn to turn but rather over a continuous homogeneous surface, the change in resistance is not a step function. Thus, devices using such resistive elements are said to have infinite resolution. These elements have inherently high resistance and poor heat dissipation and therefore can handle only low current. As with high-resistance wire-wound elements, use of an amplifier to boost the power capability of the device may be necessary.

9.4 LINEAR VARIABLE DIFFERENTIAL TRANSFORMERS

9.4.1 Basic Circuit

A linear variable differential transformer (LVDT) provides a convenient means of converting small motion into an electrical signal. A cross-section view of an LVDT is shown in Fig. 9.4. A constant

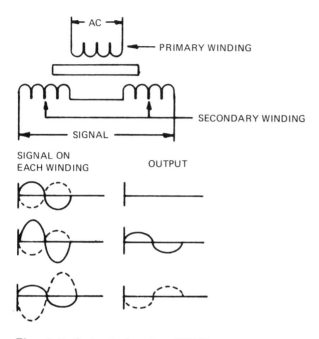

Fig. 9.5 Output signal — LVDT.

ac voltage is applied to the primary coil. The two secondary coils are identical but are connected so that they are in series opposition; that is, the signal induced in one coil is 180° out of phase with the signal of the other coil. The soft iron core passing through the center of the bobbin varies the mutual inductance, coupling the primary with the two secondary coils. If the core is positioned so that the inductive coupling between each secondary and the primary is the same, then equal voltages will be induced in each secondary. However, since the secondaries are in opposition, no output voltage will appear across the two signal leads. If the core is moved away from this position, the inductive coupling will be unbalanced and larger voltage will be induced in one secondary and a lesser voltage in the other. In this case a signal will appear across the signal leads. If the core is moved the same amount in the opposite direction, a signal will appear that is of the same magnitude but shifted 180° in phase (see Fig. 9.5).

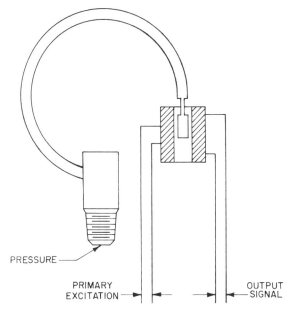

Fig. 9.6 LVDT as a transducer.

9.4.2 Performance Characteristics

LVDTs have many advantages. It is not necessary for the core to touch the i.d. of the bobbin, so there is no friction to retard the motion of the core. Good linearity and virtually infinite resolution can be attained for a core motion as small as 0.050 in. There are no wearing surfaces and the wound bobbin is not affected by shock or vibration. One disadvantage is that the primary must be excited with alternating current, preferably having a frequency on the order of 1--10 kHz. The output signal will, of course, be an ac voltage at the same frequency. With the advent of miniaturization, many transducers are built to accept dc supply voltage and to deliver dc output voltage by having integral electronics that provide the required dc/ac modulation and ac/dc demodulation.

LVDTs are commonly used for pressure transduction. The readout device (i.e., the device that converts the voltage output to a visual signal) may be digital or analog and is generally calibrated to read directly in units of pressure. Fig. 9.6 is a schematic showing the use of an LVDT as a pressure transducer.

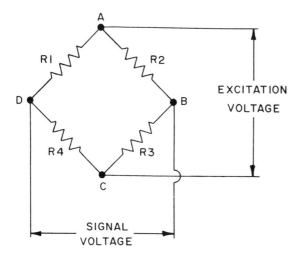

Fig. 9.7 Wheatstone bridge.

9.5 STRAIN GAUGES

9.5.1 Basic Circuit

Another transduction principle often used in pressure measurement is the strain gauge. A strain gauge element consists of fine wires or foil and exhibits a change in electrical resistance when it is strained (elastically deformed) due to an applied stress (load). By applying strain gauges to a pressure-sensitive element such as a diaphragm or a bourdon, the strain gauge may be elastically deformed in a manner proportional to pressure and thus provide a resistance change that can be measured electrically. The strain gauges may be bonded directly to the pressure-sensitive element or to a deformable member acted on by the pressure element. The most common form of strain gauge pressure transducer consists of an elastically deformable plate or beam to which are bonded four strain gauge elements (resistors) located so that deformation of the plate results in two of the elements being put in tension and two in compression. The outputs of the four elements are connected in a Wheatstone bridge circuit as shown in Fig. 9.7.

If the ratio of the resistance of R_1 to R_4 is equal to the ratio of the resistance of R_2 to R_3, then the voltage drop from point A to B will equal the voltage drop from point A to D, and there will be no potential difference between points B and D. If, however, the strain gauge elements are strained so that R_1 and R_3 increase in resistance and R_2 and R_4 decrease in resistance, then there will be a potential difference between B and D, which can be read out on a suitable analog or digital meter.

The Wheatstone bridge arrangement has several advantages. For example, the resistance of a strain gauge will change with ambient temperature. However, if all four strain gauges change resistance in the same proportion (e.g., each one increases 5% of its value), the ratio of R_1 to R_4 will remain the same as the ratio of R_2 to R_3 and there will be no output across points B and D. If a single strain gauge element were used, a change in resistance due to temperature would appear the same as one due to strain.

9.5.2 Performance Characteristics

Strain gauge pressure transducers have fast response and infinite resolution. Very little motion is required of the pressure-sensitive element, permitting the use of light but relatively stiff elements having good resistance to shock and vibration. Since there are no moving parts (other than deformation of the pressure-sensitive element), wear is not a problem. Temperature effects can be minimized by compensation. The principal disadvantage is that the output signal level is quite low, typically 3 mV per volt of excitation at full-scale pressure. This low output generally makes it necessary to incorporate an amplifier, thus adding to the cost. Because the output of a strain gauge is ratiometric; that is, for a given transducer there is a fixed ratio between the input (excitation) voltage and the output, it is necessary to precisely control the input voltage in order to obtain a known V/psi output. In most cases, strain gauges equipped with an integral amplifier to increase the level of the output signal will also be equipped with an integral voltage regulator, so the user will not have to precisely control the input voltage. A typical strain gauge pressure transducer having an integral voltage-regulated input and an amplified output might be specified as having an output of 5.0 V dc for zero to full-scale pressure when the input voltage is between 20 and 30 V dc. Note that the input and output are dc voltages.

9.5.3 Diffused Silicon Strain Gauges

A recent development in pressure transducers utilizes a silicon wafer having strain gauge elements diffused in a precise pattern

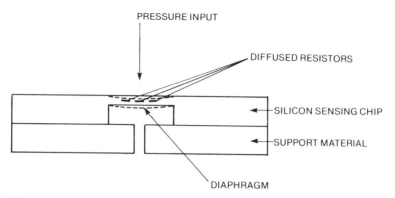

Fig. 9.8 Silicon wafer pressure sensitive element. (Courtesy of Controls Division, AMETEK, Inc.)

directly onto the silicon (see Figs. 9.8 and 9.9). Using photolithographic techniques in conjunction with the deposition of aluminum, both the contacts to the strain gauge elements and the connection pads are provided directly on the chip. A diaphragm is formed by etching away the opposite side of the wafer at a position directly under the strain gauge pattern. The size of the diaphragm may be on the order of 0.2 in. in diameter or a square 0.2 in. on a side. Thus, the transducer element becomes a monolithic structure consisting of a pressure sensitive diaphragm, strain gauge elements, electrical connections, and a support structure.

Silicon semiconductor gauges have a higher gauge factor (that is, the magnitude of the change in resistance for a given applied strain) than the bonded wire or bonded foil types and thus require less amplification to reach a given output signal level. They are highly resistant to damage by shock and vibration but are more difficult to compensate for temperature errors. Since the diaphragm is actually part of the silicon chip, the pressure medium will contact the silicon, unless a chemical separator is used. While silicon is corrosion-resistant to most industrial media, satisfactory compatibility must be considered.

9.6 CAPACITANCE TRANSDUCERS

9.6.1 Basic Considerations

Pressure transducers based on a change in capacitance are offered by several manufacturers. This type of transducer consists of two

Capacitance Transducers

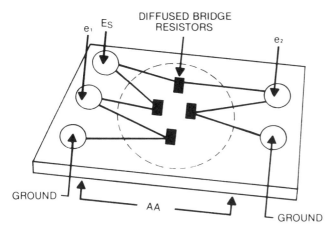

Fig. 9.9 Diffused bridge on silicon wafer. Schematic top view of a sensing chip. (Courtesy of Controls Division, AMETEK, Inc.)

parallel conducting plates spaced a small distance apart. Because the capacitance between the two conducting plates is inversely proportional to the distance between them, an uncompensated transducer would be very nonlinear. Linearizing networks are usually incorporated, but the amount of correction possible is somewhat limited. Therefore, it is necessary to keep the motion of the variable capacitor plates small in comparison to the nominal distance between the plates. Further, the air gap must be small in order to obtain a capacitance large enough to work with. As an example, if the nominal distance between the plates is 0.005 in. at zero pressure and 0.004 in. at full scale pressure (i.e., 0.001 in. motion), then the change in capacitance for a small increment of motion at full scale pressure will be approximately 0.5 times the change for the same increment at zero pressure. This degree of non-linearity is compensable. However, if the motion from zero to full scale pressure is 0.002 in. (i.e., gap at full scale pressure is 0.003 in.), then the capacitance change at full scale pressure would be approximately 2.6 times that at zero pressure, which would make compensation difficult.

Because the motion of the pressure-sensitive element used in capacitive pressure transducers is very small, dimensional changes due to temperature effects or long-term drifting are especially critical.

If the space between the two plates is filled by the pressure medium, then the pressure medium must not alter the dielectric between the plates.

9.7 ACCURACY OF PRESSURE TRANSDUCERS

9.7.1 Factors Affecting Accuracy

There are several ways to express the accuracy of pressure transducers, and when comparing specifications among various types and brands, care must be taken to be certain the comparison is being made on an equal basis. The principal performance characteristics contributing to the overall accuracy are as follows:

Linearity (scale shape)
Zero error (null output or offset)
Range error
Hysteresis
Temperature effect (zero and range)
Repeatability
Long-term stability

The last two characteristics may require some further explanation. Repeatability refers to the consistency with which the transducer will produce the same output under identical operating conditions. Long-term stability refers to the ability of the transducer to retain its accuracy over an extended period of operation or an extended period of nonuse.

9.7.2 Methods of Specifying Accuracy

A transducer having integral electronics, which condition both the power supply and the output signal, will usually have the accuracy specified as a percentage of deviation from an ideal or nominal straight line (Fig. 9.10). In this case, the user may apply the output signal directly to his system, if the impedance of the transducer and the system are appropriate. Therefore, one transducer may be simply replaced with another of the same kind without the necessity of making any adjustments to the system, and the overall accuracy of the system will remain within the original expected accuracy. In order to maintain the stated accuracy with respect to an ideal straight line, the transducer manufacturer must provide adjustments (usually in the form of variable resistor networks) to adjust for zero error, range error, and linearity. The transducer is calibrated by applying a known pressure and adjusting the resistance values to bring the output readings into tolerance. The resistors may be manually adjusted potentiometers, but more often take the form of a resistance network deposited on a substrate and trimmed to the proper value by cutting away part of the conductive

Accuracy of Pressure Transducers

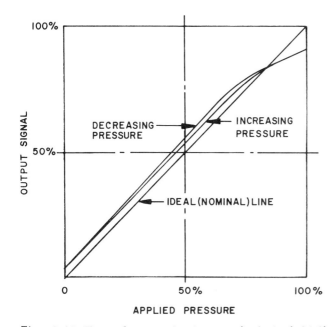

Fig. 9.10 Transducer output − nominal straight line.

path using a laser beam or abrasives. Sophisticated production equipment may be employed that automatically measures the deviations, calculates the resistance values needed for correction, and then trims to the required value.

Some users of transducers may wish to incorporate the range and zero adjustment networks as part of their own circuitry, in which case the transducer manufacturer will specify the accuracy as a percentage of deviation from the best straight line. The best straight line (BSL, or best fit straight line, or independent straight line) is a straight line positioned on a plot of the transducer output signal vs. input pressure, so that the maximum deviation of the output signal in a plus direction is equal in magnitude to the maximum deviation of the output signal in the minus direction. The slope of the line and the point at which it intercepts the zero pressure axis are chosen to obtain the equal plus and minus deviations. Fig. 9.11 shows a best straight line drawn so that the deviations A, B, and C are equal in magnitude. Specifying the accuracy in relation to a BSL

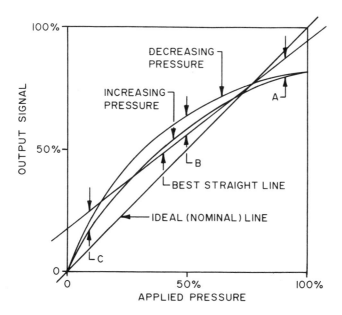

Fig. 9.11 Transducer output — best straight line.

eliminates two of the largest error sources: range error and zero error. Except for temperature errors, which will be discussed later, the only other sources of error remaining are nonlinearity, hysteresis, and lack of repeatability. Therefore, it is simpler and easier for the manufacturer to offer improved accuracy when specifying the errors to a BSL as opposed to specifying the errors to a nominal or ideal straight line. In the case of the BSL, the user then makes the required adjustments in his system electronics to standardize the zero error and range errors so as to be compatible with his system accuracy requirements. Of course, if the transducer is replaced, then the user must readjust his system electronics, since it is unlikely that the new transducer will utilize the same BSL as the one replaced.

9.7.3 Effects of Ambient Temperature

The accuracy of all types of transducers is greatly affected by ambient temperature variations. If the transducer is used in a

Accuracy of Pressure Transducers 245

temperature-controlled laboratory, the effect may not be significant, but if it is mounted in the engine compartment of a farm tractor or an aircraft, then temperature compensation is of great importance. Temperature changes affect not only the mechanical aspects of the pressure sensor (i.e., change in material modulus and dimensional variations), but also the characteristics of the various electronic components. Therefore, temperature compensation must be done by subjecting the transducer to the specified temperature range, noting the shifts, calculating the required resistor values to compensate for the shift, returning to room temperature to do the trimming operation, and then retesting over the temperature range to determine if the proper compensation was accomplished.

Obviously, temperature compensation adds considerably to the cost of the transducer. In order to keep the cost down, transducers are sometimes compensated over only a portion of the temperature range in which they are capable of operating.

9.7.4 Performance Specification

Based on all of the above, a typical transducer performance specification may be as follows:

1.	Pressure range	0–100 psi
2.	Supply voltage (excitation)	8–16 V dc
3.	Output voltage	1–5 V dc
4.	Accuracy at 70°F including hysteresis, linearity, and repeatability	±2% of span or 0.08 V
5.	Operating range	-40°F to 250°F
6.	Temperature effect over -20°F to +180°F	0.02%/°F
7.	Linearity and hysteresis	±0.5%
8.	Repeatability	0.2%
9.	Resolution	Infinite
10.	Long-term stability	±0.4%/yr
11.	Minimum load impedance	2.5 kΩ
12.	Supply current	10 mA

Other specification characteristics that may be of interest are:

1. Transient voltage protection
2. Reversed polarity protection
3. Life in terms of pressure cycles
4. Corrosion resistance

5. Proof pressure
6. Rupture pressure
7. Null offset (i.e., voltage output including polarity, when applied pressure is zero)
8. Response time
9. Natural frequency
10. Threshold sensitivity
11. Storage temperature extremes

10
Safety

10.1 INTRODUCTION

Safety is a very important consideration in the selection, installation, and use of pressure gauges. Failure to properly evaluate and guard against certain hazards can result in anything from a minor inconvenience to loss of life and severe property damage. Throughout the previous sections many references have been made to safety considerations, and it is the purpose of this chapter to gather these references together and emphasize them.

In terms of importance, this chapter might will have been the first chapter in the handbook. However, the significance of the material will be more appreciated following the background provided by the earlier chapters.

Section 4 of the American National Standards Institute Standard B40.1-1980 (see Sec. 2.1) entitled "Safety" and a portion of Section 3 entitled "General Recommendations" present important information on safety. By permission of the American Society of Mechanical Engineers this information is presented here in its entirety. It is strongly recommended that a copy of the complete standard be obtained and read.

Additional safety-related information is given in Secs. 10.4 and 10.5 courtesy of Ametek, Inc., U.S. Gauge Division.

Every user of pressure gauges should read and understand the information presented in this chapter.

10.2 EXCERPTS FROM ANSI B40.1—1980

3.3.4 *Pressure Connection*

Location of Connection

Stem Mounted — Bottom or Back
Surface Mounted — Bottom or Back
Flush Mounted — Back

Type of Connection. Taper pipe connections for pressures up through 20,000 psi are usually 1/8–27, 1/4–18, or 1/2–14 American Standard external or internal taper pipe threads (NPT) per ANSI B2.1 as required. Above this pressure 1/4 high pressure tubing connections, or equal, shall be used.

In applications of *stem mounted gauges*, especially with liquid-filled cases and where vibration is extremely severe, consideration should be given to the possibility of failure of the stem or associated piping, caused by the vibrating mass of the gauge. A larger connection (e.g., 1/2 NPT instead of 1/4 NPT) and/or a stronger stem material (e.g., stainless steel instead of brass) should be considered.

3.4.3.10 Mounting a pressure gauge in a *position* other than that at which it was calibrated can affect its accuracy. Normal calibrating position is upright/vertical. For applications requiring mounting in other than this position, notify the supplier.

3.4.3.11 CAUTION TO USERS. PRESSURE GAUGES CAN BE RENDERED INACCURATE DURING SHIPMENT DESPITE CARE TAKEN IN PACKAGING. TO INSURE CONFORMANCE TO THE STANDARD GRADE TO WHICH THE PRESSURE GAUGE WAS MANUFACTURED, IT SHOULD BE CHECKED BEFORE USE.

4.0 SAFETY

4.1 Scope

This section of the standard presents certain information to guide users, suppliers and manufacturers toward minimizing the hazards that could result from misuse or misapplication of pressure gauges with elastic elements. The user should become familiar with all sections of this standard as all aspects of safety cannot be covered in this section. Consult the manufacturer or supplier for advice whenever there is uncertainty about the safe application of a pressure gauge.

Excerpts from ANSI B40.1—1980 249

4.2 General Discussion

4.2.1 Adequate safety results from intelligent planning and careful selection and installation of gauges into a pressure system. The user should inform the supplier of all conditions pertinent to the application and environment so that the supplier can recommend the most suitable gauge for the application.

4.2.2 The history of safety with respect to the use of pressure gauges has been excellent. Injury to personnel and damage to property have been minimal. In most instances, the cause of failure has been misuse or misapplication.

4.2.3 The pressure sensing element in most gauges is subjected to high internal stresses, and applications exist where the possibility of catastrophic failure is present. Pressure regulators, chemical (diaphragm) seals, pulsation dampers or snubbers, syphons, etc., are available for use in these potentially hazardous systems.

4.2.4 CAUTION. PRESSURES IN EXCESS OF FULL SCALE PRESSURE OR VACUUM SHOULD BE AVOIDED. OVERPRESSURE MAY CAUSE CALIBRATION CHANGE, DAMAGE TO THE ELASTIC ELEMENT OR GAUGE FAILURE AND THEREFORE SHOULD NOT BE APPLIED UNLESS RECOMMENDED BY THE GAUGE MANUFACTURER.

4.2.5 The following systems are considered potentially hazardous and must be carefully evaluated:
1. Compressed gas systems.
2. Oxygen systems.
3. Systems containing hydrogen or free hydrogen atoms.
4. Corrosive fluid systems (gas and liquid).
5. Pressure systems containing any explosive or flammable mixture or medium.
6. Steam systems.
7. Non-steady pressure systems.
8. Systems where high overpressure could be accidentally applied.
9. Systems wherein interchangeability of gauges could result in hazardous internal contamination or where lower pressure gauges could be installed in higher pressure systems.
10. Systems containing radioactive or toxic fluids (liquids or gases).
11. Systems installed in a hazardous environment.

4.2.6 When gauges are to be used in contact with media having known or uncertain corrosive effects or known to be radioactive, random or unique destructive phenomena can occur. In such cases the user should always furnish the supplier or manufacturer with information relative to the application and solicit his advice prior to installation of the gauge.

4.2.7 Fire and explosions within a pressure system can cause pressure element failure with very violent effects, even to the point of completely disintegrating or melting the pressure gauge. Violent effects are also produced when failure occurs due to (1) hydrogen embrittlement, (2) contamination of a compressed gas, (3) formation of acetylides, (4) melting of soft solder joints by steam or other heat sources, (5) corrosion, (6) fatigue, (7) the presence of shock, and (8) excessive vibration. Nearly any failure in a compressed gas system produces violent effects.

4.2.8 *Modes of Elastic Element Failure*. There are four basic modes of elastic element failure: 1) failure due to fatigue, 2) failure due to application of overpressure, 3) corrosion failure and 4) explosive failure due to chemical reaction (explosion) within the element.

4.2.8.1 *Fatigue Failure*. Fatigue failure generally occurs along a highly stressed edge radius appearing as a small crack which propagates along the edge radius. Such failures are usually more critical with compressed gas media than with liquid media.

Fatigue cracks usually release the media fluid slowly so case pressure build-up can be averted by providing pressure relief openings in the gauge case. However, in high pressure elastic elements where the yield strength approaches the ultimate strength of the element material, fatigue failure may resemble explosive failure.

A restrictor placed in the gauge pressure inlet will reduce pressure surges and restrict fluid flow into the partially open bourdon tube. A restrictor should be considered for gauges used in compressed gas applications.

4.2.8.2 *Overpressure Failure*. Overpressure failure is caused by the application of internal pressure greater than the rated limits of the elastic element and can occur when a low pressure gauge is installed in a high pressure port or system. The effects of overpressure failure, usually more critical in compressed gas systems than in liquid filled systems, are unpredictable and may cause parts to be propelled in any direction. Cases with pressure relief openings will not always retain expelled parts.

Placing a restrictor in the pressure gauge inlet will not reduce the immediate effect of failure but will help control flow of escaping fluid following rupture and reduce potential of secondary effects.

It is generally accepted that solid front cases with pressure relief back will reduce the possibility of parts being projected forward in the event of failure.

The window material alone will not provide adequate protection against internal case pressure build-up and can, in fact, be the most hazardous component.

4.2.8.3 *Corrosion Failure*. Corrosion failure occurs when the elastic element has been weakened through attack by corrosive chemicals present in either the media inside or the environment outside it. Failure may occur as pinhole leakage through the element walls or early fatigue failure due to stress cracking brought about by chemical deterioration or embrittlement of the material.

A chemical (diaphragm) seal should be considered for use with pressure media that may have a corrosive effect on the elastic element.

4.2.8.4 *Explosive Failure*. Explosive failure is caused by the release of explosive energy generated by a chemical reaction such as can result when adiabatic compression of oxygen occurs in the presence of hydrocarbons. It is generally accepted that there is no known means of predicting the magnitude or effects of this type of failure. For this mode of failure a solid wall or partition between the elastic element and the window will not necessarily prevent parts being projected forward.

4.2.9 *Pressure connection*. See recommendations in paragraph 3.3.4.

4.3 Safety Recommendations

4.3.1 *Operating Pressure*. The pressure gauge selected should have a range such that the operating pressure occurs in the middle half (25 to 75%) of the scale. A good rule of thumb is to select a gauge with a full scale pressure two times the intended operating pressure.

Should it be necessary for the operating pressure to exceed 75% of full scale, contact the supplier for recommendations.

This does not apply to Test, Retarded, or Suppressed Scale gauges.

4.3.2 *Pressure Element*

4.3.2.1 *Compatibility with the Pressure Medium*. The elastic element is generally a thin-walled member which of necessity operates under high stress conditions and must, therefore, be carefully selected for compatibility with the pressure medium being measured. None of the common element materials is impervious to every type of chemical attack. The potential for corrosive attack is established by many factors including the concentration, temperature and contamination of the medium. The user should inform the gauge supplier of the installation conditions so that the appropriate element materials can be selected.

4.3.2.2 In addition to the factors discussed above, the capability of a pressure element is influenced by the design, materials and fabrication of the *joints between its parts*.

Common methods of joining are soft soldering, silver brazing, and welding. Joints can be affected by temperature, stress and corrosive media. Where application questions arise, these factors should be considered and discussed by the user and manufacturer.

4.3.2.3 Some special applications require the pressure element assembly have a high degree of *leakage integrity*. Special arrangement should be made between manufacturer and user to assure that the allowable leakage rate is not exceeded.

4.3.3 *Cases*

4.3.3.1 *Cases, Solid Front*. It is generally accepted that a solid front case per Section 3.3.1 will reduce the possibility of parts being projected forward in the event of elastic element failure. An exception is explosive failure of the elastic element.

4.3.3.2 *Cases, Liquid Filled*. It has been general practice to use glycerine or silicone filling liquids. However, these fluids may not be suitable for all applications. They should be avoided where strong oxidizing agents including, but not limited to, oxygen, chlorine, nitric acid and hydrogen peroxide are involved. In the presence of oxidizing agents potential hazard can result from chemical reaction, ignition or explosion. Completely fluorinated and/or chlorinated fluids may be suitable for such applications.

Excerpts from ANSI B40.1—1980

The user shall furnish detailed information relative to the application of gauges having liquid filled cases and solicit the advice of the gauge supplier prior to installation.

Consideration should also be given to the instantaneous hydraulic effect which may be created by one of the modes of failure outlined in 4.2.8. The hydraulic effect due to pressure element failure could cause the window to be projected forward even when a case having a solid front is employed.

4.3.4 *Restrictor*. Placing a restrictor between the pressure connection and the elastic element will not reduce the immediate effect of failure but will help control flow of escaping fluid following rupture and reduce the potential of secondary effects.

4.3.5 *Specific Service Conditions*

4.3.5.1 *Specific Applications* for pressure gauges exist where hazards are known. In most instances requirements for design, construction and use of gauges for these applications are specified by State or Federal Agencies or Underwriters' Laboratories, Inc. Some of these Specific Service gauges are listed below. The list is not intended to include all types and the user should always advise the supplier of all application details.

4.3.5.2 *Oxygen Gauge*. A gauge designed to indicate oxygen pressure. Cleanliness shall comply with Level IV (see Section 5.0). The dial shall be clearly marked with a univeral symbol and/or USE NO OIL in red color.

4.3.5.3 *Ammonia Gauge*. A gauge designed to indicate ammonia pressure and to withstand the corrosive effects of ammonia. The gauge may bear the inscription AMMONIA on the dial. It may also include the equivalent temperature scale markings on the dial.

4.3.5.4 *Acetylene Gauge*. A gauge designed to indicate acetylene pressure. It shall be constructed using materials that are compatible with commercially available acetylene. The gauge may bear the inscription ACETYLENE on the dial.

4.3.5.5 *Chemical Gauge*. A gauge designed to indicate the pressure of corrosive and/or high viscosity fluids. The primary material(s) in contact with the pressure medium may be identified on the dial. It may be equipped with a chemical seal, pulsation damper, and/or pressure relief device. These devices help to minimize potential damage to personnel and property in the event of gauge failure. They may also reduce accuracy and/or sensitivity, however.

4.4 Reuse of Pressure Gauges

It is not recommended that pressure gauges be moved from one application to another. Should it be necessary, however, the following must be considered:

4.4.1 *Chemical Compatability*. The consequences of incompatability can range from contamination to explosive failure. For example, moving an oil service gauge to oxygen service can result in explosive failure.

4.4.2 *Partial Fatigue*. The first application may involve pressure pulsation which has expended most of the gauge life, resulting in early fatigue in the second application.

4.4.3 *Corrosion*. Corrosion of the pressure element assembly in the first application may be sufficient to cause early failure in the second application.

4.4.4 *Other Considerations*. When re-using a gauge all guidelines covered in this standard relative to application of gauges should be followed in the same manner as when a new gauge is selected.

10.3 ADDITIONAL SAFETY INFORMATION

Additional important information related to safety is given in the following sections, 10.4 and 10.5.

10.4 FAILURE MODES

A very important aspect of selecting and installing pressure gauges is the consideration of the hazards that will result in the event the gauge fails.

The primary causes of failure are misapplication and/or abuse of the gauge. Those people who are responsible for the selection and installation of pressure gauges must recognize conditions that will adversely affect the ability of the gauge to perform its function or that

Factors to be Considered 255

will lead to early failure. These conditions may then be discussed with the manufacturer to obtain his recommendations. Failure may constitute:

1. Loss of accuracy
2. Clogging of the pressure port or damage to the internal mechanism so that there is either no indication when pressure is applied or an indication of pressure even though none is applied
3. A leak in the pressure-containing parts or joints
4. A crack or fatigue failure of the bourdon
5. Bursting of the bourdon due to severe overpressure
6. An explosion within the system due to a chemical reaction of the pressure medium with contaminants causing the bourdon to explode

10.5 FACTORS TO BE CONSIDERED

When specifying, using or installing a pressure gauge, the following factors must be given consideration.

10.5.1 Operating Pressure

Do not continuously operate the gauge at more than 75% of the span. Bourdons are necessarily highly stressed, especially in ranges over 1000 psi, and continuous operation at full scale will result in early fatigue failure and subsequent rupture.

10.5.2 Materials

Be certain the materials of the pressure-containing portions of the gauge are compatible with the pressurized medium. Gauges are commonly made of copper alloys (brass, bronze, etc.) and may be subject to stress corrosion or chemical attack. Bourdons have relatively thin walls, and the accuracy of the indication is directly affected by any reduction in the wall thickness. Use of the same material for the bourdon as used for the tank or associated piping is not necessarily good practice. A material having a corrosion rate of 0.001 in. year may be suitable for the piping but will be entirely unsuitable for a bourdon having a wall thickness of, for example, 0.008 in. It is imperative that the proper bourdon material be selected for the service on which the gauge is used. Gauges specially constructed for corrosion service are available.

10.5.3 Cyclic Pressure and Vibration

Continuous, rapid pointer motion will result in excessive wear of the internal mechanism and cause gross errors in the pressure indicated and possibly early fatigue failure of the bourdon. If the pointer motion is due to mechanical vibration, the gauge must be remotely mounted on a nonvibrating surface and connected to the apparatus by flexible tubing. If the pointer motion is due to pressure pulsations, a suitable damper must be used between the pressure source and the gauge.

10.5.4 Accuracy at Low-Pressure End of Span

Pressure gauges should not be used to measure pressures less than 10% of the span. The accuracy of a pressure gauge is normally stated as "percent of span." The accuracy of commercial Grade B gauges (see Sec. 2.7 for various grades of accuracy) is ±3% of span over the first quarter of the scale. If, for example, a 0–100 psi Grade B gauge is used to measure 6 psi, the accuracy of measurement will be ±3 psi or ±50% of the *applied* pressure. In addition, the scale of pressure gauges is often laid out with a "take-up" (start), which can result in further inaccuracies when measuring pressures that are a small percentage of the gauge span (see Sec. 3.13.4 for an explanation of take-up). For the same reasons, gauges should not be used for the purpose of indicating that the pressure in a tank, autoclave, etc., has been completely exhausted to atmospheric pressure. Depending on the accuracy and the span of the gauge and the possibility that a take-up is incorporated at the beginning of the scale, hazardous pressure may remain in the tank even though the gauge is indicating zero pressure. A vent valve must be used to completely reduce the pressure before unlocking covers, removing fittings, etc.

10.5.5 Fatigue

As with any spring, the bourdon will fail after extended use and release the pressurized medium. The larger the number of applied pressure cycles and the greater the extent of the pressure cycles, the earlier failure will occur. The fatigue failure may be explosive. Since such a failure will be hazardous to personnel or property, precautions must be taken to contain or direct the release of the pressurized medium in a safe manner.

10.5.6 Frequency of Accuracy Evaluation

Where the pressure measurement is critical and gauge failure or gross inaccuracy will result in a hazard to personnel or property, the gauge

Factors to be Considered 257

should be checked for accuracy and proper operation on a periodic basis.

10.5.7 Use with Oxygen

Gauges used for measurement of oxygen pressure must be free of contamination within the pressure-containing portion. Various levels of cleanliness are specified in ANSI B40.1. The gauge itself and the equipment to which the gauge is attached (pressure regulators, cylinders, etc.) must be kept clean so as not to contaminate the gauge. Filters on the equipment must be examined periodically and cleaned or replaced. The sudden inrush of a high-pressure gas will momentarily create a very high temperature which in the presence of oxygen may ignite the contaminant, causing a violent explosion. *Therefore, when the valve on the oxygen supply tank is opened to admit oxygen to the regulator, the valve should be opened very slowly so as to allow the pressure to build up slowly.* In order to accomplish this it is recommended that the tank valve be opened momentarily and then closed snugly but not excessively before attaching the regulator. This will not only blow out accumulated dirt in the valve but will also place the valve in a condition that will permit it to be opened slowly rather than suddenly break loose as a result of being closed too tightly. When bleeding the oxygen tank prior to attaching the regulator, be certain that the valve opening is directed away from any open flame and the operator. *When opening the oxygen tank valve, the operator must not stand in front of or behind the gauge and must wear eye and face protection.* In this position, if there is an explosion due to contaminated equipment, any particles projected from the gauge will not be propelled directly at the operator.

10.5.8 Use with Hydrogen

Steel bourdons, including 400 series stainless steel, are subject to hydrogen embrittlement when stressed. Measurement of gas or liquids containing hydrogen (such as natural gas and sour oil) requires the use of special materials for the bourdon.

10.5.9 Venting of Case

Vents provided in the pressure gauge case (clearance around pressure connection, rubber grommets, pressure-relief back, etc.) must not be closed or restricted from operating. There is always the possibility that the pressure medium will be admitted to the case interior as a result of a leaking joint or bourdon tube failure. If this occurs, the pressure medium must be vented from the case so as not to build up sufficient pressure to rupture the case or window. *However, venting will not prevent case rupture in the event of a violent explosion.*

10.5.10 Liquid-Filled Gauges

Performance of pressure gauges used in severe vibration or pulsating pressure service can be improved by filling the gauge case with a viscous fluid. Gauges constructed in this manner necessarily require sealed cases to prevent the escape of the liquid. However, some means of venting the case must be provided. In some instances, this vent is sealed to prevent loss of fluid during shipment and must be released after the gauge is installed. Be certain to follow the installation instructions for properly venting the gauge after installation. The liquid filling most commonly used is a mixture of glycerin and water. *Glycerin can combine with strong oxidizing agents including, but not limited to, chlorine, nitric acid, and hydrogen peroxide and result in an explosion that can cause property damage and personal injury. If gauges are to be used in such service, do not use glycerin- or silicone-filled gauges; consult the factory for proper filling medium.*

11
Metric Considerations

11.1 INTRODUCTION

Sec. 1.7 discusses the metric or SI system of pressure units (Le Système International d'Unites), which was adopted in 1960 at the General Conference on Weights and Measures (CGPM). In accordance with this system, the pascal, together with metric prefixes such as kilo (k) and mega (M), is the only recognized unit of pressure. However, there is strong support for continued use of another pressure unit, the bar, which has been in use for many years.

11.1.1 Bar as a Unit of Pressure

The bar is not a gravimetric unit; that is, its definition does not include a value for the magnitude of the earth's gravity. It is derived from another metric system (CGS) using the centimeter as the unit of length, the gram as the unit of mass, and the second as the unit of time. The unit of force in this system is called a dyne and is defined as the force required to accelerate a mass of 1 g at the rate of 1 cm/sec^2. The unit of pressure in this system becomes a dyn/cm^2. One million dyn/cm^2 is very nearly 1 standard atm. As a matter of convenience, the bar was created and is defined as 1 million dyn/cm^2, which is equivalent to 14.504 psi.

The unit pascal represents a very small pressure which, for general use, is not practical. Even the kilopascal is small, being equal to only 0.1450 psi. For hydraulic power applications, working pressures of 20,000 kPa (2903 psi) are common. Such a large number is inconvenient to handle and more subject to error in reading and recording. Since the parameters in a bar (grams, seconds, centimeters) are decimal multiples of the parameters in a pascal (kilograms, seconds, and meters), the relationship between the two units is a decimal multiple and 1 bar equals 100 kPa exactly. Therefore the working pressure in the hydraulic power example above may be expressed as 200 bars.

An alternative solution to the use of bar for high-pressure applications is to use the metric multiplier *mega-*, which means 1 million. In this case we have a megapascal (abbreviated MPa), and in the same hydraulic power example the pressure could be expressed as 20 MPa. However, there is great reluctance to manufacture gauges having dials graduated in terms of MPa because of its similarity with the more common kPa. There is the possibility that a user of the gauge, who is not sufficiently knowledgeable of the SI system, will not recognize and differentiate between the two units, with the consequence that either the gauge or the system in which the gauge is used will be catastrophically overpressured. Therefore, if *bar* is not permitted, it is best to graduate the dial in kPa and show the required number of zeroes.

11.1.2 kg/cm^2 as a Unit of Pressure

Prior to the SI system, the metric unit of pressure was the gravimetric unit kilogram per square centimeter (kg/cm^2). In this pressure unit, the kilogram is used as a unit of force, not as a unit of mass as defined in the SI system. This is an unfortunate contradiction of the original concept of the kilogram as a unit of mass. However, the use of kilogram as a unit of force is so widespread that it is unlikely that the general public will ever stop the practice. In order to clarify the situation, some technical literature will present this pressure unit as kilogram of force per square centimeter (kgf/cm^2). Actually the kilogram of force is that force developed by the earth's gravitational attraction on 1 kg of mass, in other words, the "weight" of 1 kg of mass at standard gravity. The unit of force in the SI system is the newton, which is the force required to accelerate a mass of 1 kg at the rate of 1 m/sec^2. Since the acceleration due to the earth's gravity is defined as 9.80665 m/sec^2 (the assumed standard), then a mass of 1 kg being acted on by standard gravity will result in a force of 9.80665 newtons, which is equivalent to saying a kilogram of force is equal to 9.80665 newtons. Continuing this conversion to a pressure unit

Introduction

will illustrate the relationship between the gravimetric unit, kilogram per square centimeter, and the kilopascal.

$$1 \text{ kilogram of force} = 9.80665 \text{ N}$$

$$\text{or } 1 \text{ newton} = \frac{1}{9.80665} \text{ kg}_f \qquad (1)$$

$$1 \text{ meter} = 100 \text{ cm} \qquad (2)$$

$$1 \text{ kilopascal} = 1000 \times \frac{N}{m^2} \qquad (3)$$

Substituting the values of Eq. (1) and Eq. (2) in Eq. (3) gives

$$1 \text{ kPa} = 1000 \times \frac{\text{kg}_f}{9.80665} \times \frac{1}{(100 \text{ cm})^2} \qquad (4)$$

$$\frac{1 \text{ kg}_f}{\text{cm}^2} = 98.0665 \text{ kPa (exactly)}$$

11.1.3 Conversion of Kilopascal to Psi

When converting kilopascal into English units, it is necessary to know the relationship between 1 kg of mass and 1 lb of mass, which is the exact numerical equivalent of the relationship between 1 kg of force and 1 lb of force. The United States National Bureau of Standards publication LC1035 (January 1976) establishes this value at 0.453 592 37 kg of mass equals 1 lb of mass exactly. This same publication establishes the relationship between centimeters and inches as 2.54 cm equals 1 in. exactly. Substituting these equivalents of the metric quantities in Eq. (4) gives

$$1 \text{ kPa} = 1000 \times \frac{1}{0.453\ 592\ 37} \times \frac{1}{9.80665} \times \frac{1}{(100/2.54)^2}$$

$$= 0.145\ 037\ 737\ 7 \text{ psi}$$

11.1.4 Kilopond/cm² as a Unit of Pressure

Another metric unit of measure common in some European countries is the kilopond per square centimeter (kp/cm^2). The unit is based on a unit of force designated the pond, which is the force developed by the earth's standard gravitational force acting on a mass of 1 g. A kilopond, then, is the force developed by a kilogram of mass in the earth's field, and therefore a kilopond per square centimeter (kp/cm^2) is exactly equal to a kilogram of force per square centimeter (kg_f/cm^2). Since this is a gravimetric pressure unit, its use is not sanctioned by the SI system. A kp/cm^2 is called a technical atmosphere and is equal to 14.2233 psi.

Introduction of the pond as a unit of force only adds to the already existing confusion. Its spelling is so close to the English *pound* that it might be interpreted as a misspelling. In addition, there is an English unit called a kip, which is equivalent to 1000 psi and is generally abbreviated as k. The abbreviation for kilopond, kp, should not be confused with the kip.

11.1.5 Torr as a Unit of Pressure

The pressure unit torr is defined as 1/760 of an atmosphere, which, for practical purposes, makes a torr equal to one millimeter of mercury (1 mm Hg). The torr is usually used in conjunction with the measurement of vacuum and, while not defined as such, it implies absolute pressure. For example, a pressure of 10 torr means a pressure of 10 mm Hg above absolute zero.

11.2 USE OF GRAVIMETRIC UNITS

Gravimetric pressure units such as kg/cm^2, psi, and in. H_2O are so deeply ingrained among general users of pressure gauges that their use will likely continue for some time. Acceptance of any of the metric pressure units in the United States has been slow, so the great majority of the pressure gauges manufactured for general use are calibrated in psi. This has been in part due to the general resistance to any change and in part due to the refusal of the International Group to sanction the use of the bar.

In an attempt to acquaint the public with the SI system many dials are made with dual pressure scales, showing both psi and kPa or psi and bar.

11.3 NEGATIVE PRESSURE

In conjunction with the revision to existing standards made necessary by the promulgation of the SI system of units, many standard writing bodies took the opportunity to introduce the concept of negative pressure. While the concept is just as applicable to gauges calibrated in non-SI units, its use seems to be principally with gauges calibrated in SI units (pascal). Further discussion of negative pressure follows.

For many years gauges measuring vacuum (that is, pressure below the surrounding atmospheric pressure or ambient pressure) were constructed so that the pointer rotated clockwise for increasing vacuum. The dial was marked "vacuum" and the pressure unit shown was inches of mercury. However, the International Standards Organization has recommended that pressure less than ambient pressure (usually atmospheric) be referred to as negative pressure rather than vacuum. By definition, negative pressure is gauge pressure less than ambient. This concept eliminates the possible confusion that can exist when someone says he has a "vacuum of 10 inches." While such a statement should be interpreted as meaning the pressure is 10 in. of mercury below atmospheric, it could be interpreted as meaning the pressure is 10 in. of mercury above absolute zero, or approximately 19.9 in. of mercury below atmospheric. To further emphasize the negative concept, it is also recommended that the pointer move counterclockwise for increasing negative pressure and that a minus sign (-) be placed in front of the numerals. Thus, a dial for use with a gauge measuring negative pressure would appear as shown in Fig. 2.2c. The minus sign indicates the pressure is negative so, in the same manner as with positive-pressure gauges, it is not necessary to show the word "pressure" on the dial, although it would not be wrong to do so. Of course it would be in error to show the words "negative pressure" *and* the minus sign, since this is, in effect, a double negative. The use of negative pressure was adopted in many European countries, but in the United States the acceptance has not been as rapid; many users still prefer to show the word "vacuum" on the dial, have the pointer move clockwise for increasing vacuum, and not use minus signs to indicate negative pressure. When the word vacuum is shown on the dial, it would be incorrect to include the minus sign in front of the numerals, since, again, this is in effect a double negative. The above discussion on the usage of the minus sign and the word "vacuum" holds true regardless of the direction in which the pointer moves.

11.4 FOREIGN STANDARDS

In an effort to coordinate the pressure standard existing in individual countries (such as the ANSI B 40.1 standard in the United States), a technical group has been formed within the International

Standards Organization (ISO). As might be expected of such an undertaking, progress is slow due to the many differences of opinion and the many national groups that must be brought into agreement. It is necessary to keep abreast of progress made by this group.

Germany has a very strong national standards organization called the Deutsches Institut fur Normung e V, which has formulated standards on a wide variety of processes and products, including pressure gauges. Standards issued by this group are referred to as DIN standards and are widely used in European countries. Many have been officially translated into English. Translations of those that have not (including several related to pressure gauges) must be made very carefully by technically oriented people to avoid misinterpretation.

In general, the national standards of European countries are very detailed with respect to dimensions; in some cases they even specify the shape of the pointer. Gauges commonly made in the United States do not meet many of these detailed dimensions and would require complete new tooling to do so.

12
Ordering and Specification Information

12.1 INTRODUCTION

From all of the foregoing discussions, it is obvious that when ordering a gauge many decisions must be made concerning what is wanted. In making these decisions it is probable that some compromise will be required between what is wanted, what can be obtained, and the cost.

Pressure-gauge requirements generally fall into two broad classes.

1. For in-plant use on a piece of equipment used in a manufacturing process. In this case, the quantity of gauges required may be as small as one and generally not more than, say, 100 at any one time.

2. For use by an original equipment manufacturer (OEM) on a product that is made for sale. Examples would be gauges for air compressors, household water system pumps, fire extinguishers, etc. In this case quantities of several hundred thousand may be purchased by the OEM.

12.2 GAUGES FOR IN-PLANT USE

Sales of gauges in the first category are most often made through one of the manufacturer's local distributors, who will have a stock of the more popular varieties. It is obvious that when purchasing gauges from a distributor it is necessary to buy standard stock gauges in order to avoid high cost and long delivery time. Therefore, the best procedure is to first study the manufacturer's catalogs. Based on information given in earlier chapters of this handbook, select one or more model numbers as candidates and then discuss the application with the distributor, who can give assistance in making a final selection.

12.3 GAUGES FOR USE BY ORIGINAL EQUIPMENT MANUFACTURERS

If the gauges to be purchased are in the second category, it is still best to use a catalog standard. However, the manufacturer will be willing to make "nonstandard" gauges. The degree of nonstandard may be simply a special color on the case, the customer's logo printed on the dial, some different combination of standard parts, or a special thread or location of the pressure connection. If the quantity is sufficient to justify the cost of designing and tooling up, an entirely new gauge can be created. In such cases it is best to contact the manufacturer's local sales office to discuss the requirement and perhaps visit the manufacturing plant for further discussion.

12.4 INFORMATION SUPPLIED TO GAUGE MANUFACTURER

Regardless of whether the order is for stock items purchased in small quantities through a distributor or a large quantity of custom-built gauges purchased from the manufacturer, it is essential to furnish all pertinent information. The material in the previous chapters will be of great assistance in determining what information is required by the manufacturer. The following is a checklist of features that must be considered when selecting a gauge. A review of Chaps. 2, 3, and 6 will be of value.

1. General field of use
   a. Commercial
   b. Industrial
   c. Process
2. Specific end use
   a. Oxygen
   b. Ammonia

c. Hydraulic
d. Receiver
e. Boiler
f. Refrigeration
g. Welding (gas pressure)
h. Sprinkler
3. Function
   a. Pressure
   b. Vacuum
   c. Compound
   d. Duplex
   e. Differential
   f. Retard
   g. Suppressed
4. Pressure medium
   a. Gas
   b. Liquid
   c. High viscosity
   d. Corrosive
   e. High temperature
   f. Contains solid matter
   g. Subject to solidifying
   h. Possibility of hydrogen embrittlement
   i. Explosive
5. Service conditions
   a. Accuracy required
   b. Pulsating pressure
   c. External vibration
   d. Temperature extremes
       Ambient
       Pressure medium
   e. Overpressure
   f. Corrosive atmosphere
   g. Outdoor exposure
       Seacoast
       Inland
   h. Mechanical shock
6. Case type
   a. Size of gauge
   b. Method of mounting
   c. Size and location of connection
   d. Ring (bezel) style
   e. Case and ring material
   f. Window style and material

      g. Finish (case and ring)
         Paint (color)
         Electroplate
7. Pressure-containing element
      a. Pressure range
         Working pressure
         Overpressure
      b. Material of bourdon, socket, tip
      c. Method of joining
         Soft solder
         Braze
         Weld
      d. Restrictor in pressure port
      e. Fatigue life
      f. Cleanliness
8. Dial
      a. Units of pressure
      b. Special nomenclature
      c. Operating zones
      d. Material
      e. Color of graduations and background
9. Movement and pointer
      a. Material
      b. Adjustable
      c. Heavy-duty
10. Accessories
      a. Maximum or minimum pointer
      b. Diaphragm seal
      c. Pulsation damper
      d. Bleeder

Appendix: Tabular Data

This is the final chapter of this manual, in which selected tabular data are given to assist in conversion between various pressure units, between metric and English units, etc. Other data such as vapor pressure and temperature relationships, altitude pressure, etc., are included.

TABLE A.1 Pressure Conversion Chart

Pressure unit	psi	kPa	bar	atm
Pounds per sq. in. (psi)	1	6.8948	0.068948	0.068046
Kilopascal (kPa)	0.14504	1	0.01[a]	0.0098692
Bar	14.504	100[a]	1	0.98692
Atmosphere (standard)	14.696	101.33	1.0133	1
Kilograms per sq. cm (kg/cm^2)	14.223	98.067	0.98067	0.96784
Ounces per sq. in. ($oz/in.^2$)	0.0625[a]	0.43092	0.0043092	0.0042529
Inches water (in. H_2O) (20°C)	0.036063	0.24864	0.0024864	0.0024539
Centimeters water (cm H_2O) (20°C)	0.014198	0.097891	0.00097891	0.00096610
Millimeters mercury (mm Hg) (0°C)	0.019337	0.13332	0.0013332	0.0013158
Inches mercury (in. Hg) (0°C)	0.49115	3.3864	0.033864	0.033421

Density of water at 20°C = 0.998207 g/cm^3; density of mercury at 0°C = 13.5950889 g/cm^3; 1 atm = 1013250 dyn/cm^2[a]; 1 lb of force = 0.45354237 kg of force[a]; standard gravity = 9.80665 m/S^2[a]; 1 in. = 2.54 cm[a]; 1 bar = 1,000,000 dyn/cm^2.[a]

[a]Exact values.

kg/cm²	oz/in.²	in. H_2O	cm H_2O	mm/Hg	in. Hg
0.070307	16[a]	27.730	70.433	51.715	2.0360
0.010197	2.3206	4.0218	10.215	7.5006	0.29530
1.0197	232.06	402.18	1021.5	750.06	29.530
1.0332	235.14	407.51	1035.1	760.00	29.921
1	227.57	394.41	1001.8	735.56	28.959
0.0043942	1	1.7331	4.4021	3.2322	0.12725
0.0025354	0.57700	1	2.54[a]	1.8650	0.073424
0.00099821	0.22717	0.39370	1	0.73424	0.028907
0.0013595	0.30939	0.53620	1.3620	1	0.039370
0.034531	7.8585	13.620	34.594	25.4[a]	1

TABLE A.2 Metric Prefixes

Multiplication factor	Prefix	Symbol
$1.000\ 000\ 000 = 10^9$	giga	G
$1.000\ 000 = 10^6$	mega	M
$1.000 = 10^3$	kilo	k
$0.001 = 10^{-3}$	milli	m
$0.000\ 001 = 10^{-6}$	micro	μ
$0.000\ 000\ 000\ 1 = 10^{-9}$	nano	n

TABLE A.3 Metric Equivalents

1 Inch	= 25.4[a] Millimeters
1 Foot	= 0.3048[a] Meters
1 Millimeter	= 0.039370 Inches
1 Meter	= 39.370 Inches
1 Meter	= 3.2808 Feet
1 Mile	= 1.6093 Kilometers
1 Kilometer	= 0.62137 Miles
1 Pound	= 0.45359 Kilograms
1 Kilogram	= 2.2046 Pounds
1 Ounce	= 0.028350 Kilograms
1 Kilogram	= 35.274 Ounce
1 Ton (short)	= 907.18 Kilograms
1 Pound-force	= 4.4482 Newtons
1 Kilogram-force	= 9.80665[a] Newtons
1 Cubic foot	= 0.028317 Cubic meters
1 Cubic meter	= 35.315 Cubic feet

TABLE A.3 (continued)

1 Cubic inch	= 16.387 Cubic centimeters
1 Gallon	= 3785 Cubic centimeters
1 Liter	= 1000[a] Cubic centimeters
1 Cubic meter	= 264.2 Gallons
1 Cubic meter	= 1000[a] Liters
1 Gallon per minute	= 0.22715 Cubic meters per hour
1 Mile per hour	= 1.6904 Kilometers per hour

[a]Exact.

TABLE A.4 English Equivalents

1 Barrel (oil)	= 42 Gallons
1 Cubic foot	= 1728 Cubic inches
1 Cubic foot	= 7.481 Gallons
1 Cubic yard	= 27 Cubic feet
1 Gallon	= 231 Cubic inches
1 Mile	= 5280 Feet
1 Gallon per minute	= 8.021 Cubic feet per hour

TABLE A.5 Useful Data

Standard gravity	= 9.80665 m/sec^2
1 Atmosphere	= 1013250 dyn/cm^2
1 Atmosphere	= 14.696 psi
Density of water @ 20°C	= 0.998207 g/cm^3 or 62.3160 lb/ft^3
Density of mercury @ 0°C	= 13.5950887 g/cm^3

TABLE A.6 Altitude Pressure

Altitude above sea level (feet)	Average atmospheric pressure	
	Inches of mercury (abs.)	Pounds per square inch (abs.)
0	29.921	14.696
500	29.400	14.440
1,000	28.879	14.184
1,500	28.357	13.928
2,000	27.836	13.672
2,500	27.315	13.416
3,000	26.831	13.178
3,500	26.348	12.941
4,000	25.864	12.703
4,500	25.381	12.466
5,000	24.897	12.228
5,500	24.449	12.008
6,000	24.001	11.788
6,500	23.552	11.568
7,000	23.104	11.348
7,500	22.656	11.128
8,000	22.241	10.924
8,500	21.826	10.720
9,000	21.411	10.516
9,500	20.996	10.312
10,000	20.581	10.108

Appendix: Tabular Data

TABLE A.6 (continued)

Altitude above sea level (feet)	Average atmospheric pressure	
	Inches of mercury (abs.)	Pounds per square inch (abs.)
10,500	20.197	9.9200
11,000	19.814	9.7318
11,500	19.430	9.5432
12,000	19.047	9.3551
12,500	18.663	9.1665
13,000	18.309	8.9926
13,500	17.955	8.8188
14,000	17.601	8.6449
14,500	17.247	8.4710
15,000	16.893	8.2972
15,500	16.567	8.1370
16,000	16.241	7.9769
16,500	15.914	7.8163
17,000	15.588	7.6562
17,500	15.262	7.4961
18,000	14.962	7.3487
18,500	14.662	7.2014
19,000	14.361	7.0535
19,500	14.061	6.9062
20,000	13.761	6.7589

Source: This table is taken from context and is part of the ARDC Model Atmosphere Table dated 1959.

TABLE A.7 Decimal Equivalents

		1/64	0.015625			17/64	0.265625
	1/32		0.03125		9/32		0.28125
		3/64	0.046875			19/64	0.296875
1/16			0.0625	5/16			0.3125
		5/64	0.078125			21/64	0.328125
	3/32		0.09375		11/32		0.34375
		7/64	0.109375			23/64	0.359375
1/8			0.125	3/8			0.375
		9/64	0.140625			25/64	0.390625
	5/32		0.15625		13/32		0.40625
		11/64	0.171875			27/64	0.421875
3/16			0.1875	7/16			0.4375
		13/64	0.203125			29/64	0.453125
	7/32		0.21875		15/32		0.46875
		15/64	0.234375			31/64	0.484375
1/4			0.25	1/2			0.5

Appendix: Tabular Data

		33/64	0.515625			49/64	0.765625
	17/32		0.53125		25/32		0.78125
		35/64	0.546875			51/64	0.796875
9/16			0.5625	13/16			0.8125
		37/64	0.578125			53/64	0.828125
	19/32		0.59375		27/32		0.84375
		39/64	0.609375			55/64	0.859375
5/8			0.625	7/8			0.875
		41/64	0.640625			57/64	0.890625
	21/32		0.65625		29/32		0.90625
		43/64	0.671875			59/64	0.921875
11/16			0.6875	15/16			0.9375
		45/64	0.703125			61/64	0.953125
	23/32		0.71875		31/32		0.96875
		47/64	0.734375			63/64	0.984375
3/4			0.75	1			1.

TABLE A.8 Saturated Steam — Pressures, Temperatures, and Specific Volumes

Absolute pressure (psi)	Temp. (°F)	Specific volume (ft³/lb)	Absolute pressure (psi)	Temp. (°F)	Specific volume (ft³/lb)
5	162.2	73.6	34	257.6	12.2
6	170.1	62.0	35	259.3	11.9
7	176.8	53.7	36	260.9	11.6
8	182.9	47.4	37	262.6	11.3
9	188.3	42.4	38	264.2	11.0
10	193.2	38.4	39	265.7	10.8
11	197.7	35.2	40	267.2	10.5
12	202.0	32.4	41	268.7	10.3
13	205.9	30.1	42	270.2	10.0
14	209.6	28.1	43	271.6	9.8
14.7	212.0	26.8	44	273.0	9.6
15	213.0	26.3	45	274.4	9.4
16	216.3	24.8	46	275.8	9.2
17	219.4	23.4	47	277.1	9.0
18	222.4	22.2	48	278.4	8.8
19	225.2	21.1	49	279.7	8.7
20	228.0	20.1	50	281.0	8.5
21	230.6	19.2	51	282.3	8.4
22	233.1	18.4	52	283.5	8.2
23	235.5	17.6	53	284.7	8.0
24	237.8	16.9	54	285.9	7.9
25	240.1	16.3	55	287.1	7.8
26	242.2	15.7	56	288.2	7.7
27	244.4	15.1	57	289.4	7.5
28	246.4	14.7	58	290.5	7.4
29	248.4	14.2	59	291.6	7.3
30	250.3	13.7	60	292.7	7.2
31	252.2	13.3	61	293.8	7.1
32	254.0	12.9	62	294.8	6.9
33	255.8	12.6	63	295.9	6.8

Appendix: Tabular Data

TABLE A.8 (continued)

Absolute pressure (psi)	Temp. (°F)	Specific volume (ft³/lb)	Absolute pressure (psi)	Temp. (°F)	Specific volume (ft³/lb)
64	296.9	6.7	94	323.4	4.7
65	298.0	6.7	95	324.1	4.6
66	299.0	6.6	96	324.9	4.6
67	300.0	6.5	97	325.6	4.6
68	301.0	6.4	98	326.3	4.5
69	302.0	6.3	99	327.1	4.5
70	302.9	6.2	100	327.8	4.4
71	303.9	6.1	101	328.5	4.4
72	304.8	6.0	102	329.2	4.3
73	305.8	6.0	103	330.0	4.3
74	306.7	5.9	104	330.7	4.3
75	307.6	5.8	105	331.4	4.2
76	308.5	5.7	106	332.0	4.2
77	309.4	5.7	107	332.7	4.2
78	310.3	5.6	108	333.4	4.1
79	311.2	5.5	109	334.1	4.1
80	312.0	5.5	110	334.8	4.0
81	312.9	5.4	111	335.4	4.0
82	313.7	5.3	112	336.1	4.0
83	314.6	5.3	113	336.8	3.9
84	315.4	5.2	114	337.4	3.9
85	316.2	5.2	115	338.1	3.9
86	317.1	5.1	116	338.7	3.8
87	317.9	5.1	117	339.4	3.8
88	318.7	5.0	118	340.0	3.8
89	319.5	4.9	119	340.6	3.8
90	320.5	4.9	120	341.2	3.7
91	321.1	4.8	121	341.9	3.7
92	321.8	4.8	122	342.5	3.7
93	322.6	4.7	123	343.1	3.6

TABLE A.8 (continued)

Absolute pressure (psi)	Temp. (°F)	Specific volume (ft³/lb)	Absolute pressure (psi)	Temp. (°F)	Specific volume (ft³/lb)
124	343.7	3.6	158	362.5	2.9
125	344.3	3.6	160	363.5	2.8
126	344.9	3.6	162	364.5	2.8
127	345.5	3.5	164	365.5	2.8
128	346.1	3.5	166	366.5	2.7
129	346.7	3.5	168	367.4	2.7
130	347.3	3.5	170	368.4	2.7
131	347.9	3.4	172	369.3	2.6
132	348.5	3.4	174	370.3	2.6
133	349.1	3.4	176	371.2	2.6
134	349.6	3.4	178	372.1	2.6
135	350.2	3.3	180	373.1	2.5
136	350.8	3.3	182	374.0	2.5
137	351.3	3.3	184	374.9	2.5
138	351.9	3.3	186	375.7	2.5
139	352.5	3.2	188	376.6	2.4
140	353.0	3.2	190	377.5	2.4
141	353.6	3.2	192	378.4	2.4
142	354.1	3.2	194	379.2	2.4
143	354.7	3.2	196	380.1	2.3
144	355.2	3.1	198	380.9	2.3
145	355.8	3.1	200	381.8	2.3
146	356.3	3.1	205	383.9	2.2
147	356.8	3.1	210	385.9	2.2
148	357.4	3.0	215	387.9	2.1
149	357.9	3.0	220	389.9	2.1
150	358.4	3.0	225	391.8	2.0
152	359.5	3.0	230	393.7	2.0
154	360.5	2.9	235	395.5	2.0
156	361.5	2.9	240	397.4	1.9
			245	399.2	1.9

Appendix: Tabular Data

TABLE A.8 (continued)

Absolute pressure (psi)	Temp. (°F)	Specific volume (ft³/lb)	Absolute pressure (psi)	Temp. (°F)	Specific volume (ft³/lb)
250	401.0	1.8	360	434.4	1.3
255	402.7	1.8	370	437.0	1.3
260	404.4	1.8	380	439.6	1.2
265	406.1	1.7	390	442.1	1.2
270	407.8	1.7	400	444.6	1.2
275	409.4	1.7	410	447.0	1.1
280	411.0	1.6	420	449.4	1.1
285	412.6	1.6	430	451.7	1.1
290	414.2	1.6	440	454.0	1.1
295	415.8	1.6	450	456.3	1.0
300	417.3	1.5	460	458.5	1.0
310	420.3	1.5	470	460.7	1.0
320	423.3	1.4	480	462.8	1.0
330	426.2	1.4	490	464.9	0.9
340	429.0	1.4	500	467.0	0.9
350	431.7	1.3			

TABLE A.9 Vapor Pressure vs. Temperature — Ethyl Ether

Temperature (°F)	Pressure (psi)[a]	Temperature (°F)	Pressure (psi)[a]
100	4.7	250	139.1
125	13.8	275	189.6
150	26.9	300	252.0
175	45.5	325	325.4
200	69.9	350	410.7
225	100.5	375	508.5

[a]Atmospheric pressure assumed to be 14.7 psia.

TABLE A.10 Vapor Pressure Versus Temperature — Propane

Temperature (°F)	Pressure (psi)[a]	Temperature (°F)	Pressure (psi)[a]
-40	4.5	100	173.0
-20	13.0	120	226.3
0	25.0	140	291.5
20	42.4	160	367.9
40	65.6	180	455.9
60	94.6	200	558.5
80	129.8		

[a] Atmospheric pressure assumed to be 14.7 psia.

TABLE A.11 Method of Converting Temperature

Fahrenheit to Celsius (centigrade)

    Subtract 32 from Fahrenheit value
    Divide by 9
    Multiply by 5

    Example: 212°F $\dfrac{(212-32) \times 5}{9} = 100°C$

    Example: -20°F $\dfrac{(-20 - 32) \times 5}{9} = -28.89°C$

Celsius to Fahrenheit

    Divide Celsius value by 5
    Multiply by 9
    Add 32

    Example: 100°C $\dfrac{100 \times 9}{5} + 32 = 212°F$

    Example: -10°C $\dfrac{-10 \times 9}{5} + 32 = 14°F$

Celsius to Kelvin (° absolute)

    Add 273.2 to Celsius value

Fahrenheit to Rankine (° absolute)

    Add 459.7 to Fahrenheit value

Index

Accuracy:
  conditions affecting, 183
  definition of, 32
  effect of shipping on, 248
  frequency of evaluation for, 256
  grades of, 34-36
  at low end of scale, 256
  methods of expressing, 32
  as percent of indicated reading, 32-33
  as percent of span, 32-33
  selection of, 181
  tapping prior to reading, 180, 196
  of test standards, 187
Adjustment:
  determining required, 206-209
  for hysteresis, 201
  for linearity, 202
  for range, 82-84, 201
  for scale shape, 202
  for zero shift, 205
Admiralty metal, properties of, 61
Alloy steel, properties of, 64
Altitude:
  effect on pressure measurement, 14
  effect on thermometers, 227
Ambient pressure:
  effect on pressure measurement, 13
American National Standards Institute (ANSI) 20, 34, 247
American Society of Mechanical Engineers (ASME), 20, 247
Aneroid barometer, 13
ANSI (*see* American National Standards Institute)
Arbor, 80
ASME (*see* American Society of Mechanical Engineers)
Assembly of bourdon element, 74-77
  methods of, 252
    silver brazing, 75
    soft solder, 75
    welding, 76
Automotive transducers, 230-233

283

Bar:
  definition of, 15, 259
Barometer, 12
  aneroid, 13, 47
  torricellian, 12
Bar stock, 54
Bellows, 37, 44, 68-70
  characteristics, 69
  materials for, 44
  spring loaded, 69
  use in pressure gauge, 68
  welded, 70
Beryllium copper:
  properties of, 61
Best straight line, 243
Bimetal:
  as temperature compensation link, 90, 217
Bleeders, 152-155
  use in:
    filling, 153
    flushing, 153
    head correction, 153
Bourdon:
  assembly of, 74-77
  design factors, 55
  linearity of, 58
  materials for, 55, 56, 61
  motion, reason for, 40, 56
  overpressure of, 60, 81
Bourdon, C-type, 37, 38-41
  cross section, 60
  design factors, 55, 60
  minimum span, 41
  tip travel, 40, 56, 59
Bourdon gauge:
  geometry of layout, 58
  operating principle, 40
Bourdon, helical, 37, 42-44, 60
Bourdon, spiral, 37, 41

Calibration, 206-208
  correction, recording of, 197
  data sheet, 198
  definition of, 184
  error, recording of, 197
  frequency of, 183, 256
  of pressure gauges, 196-199
  prior to use, 248
  reading methods, 196
  record of readings, 198
  safe operation during:
    use of pressure regulator, 185
    use of protective shield, 185
  test stand, 188
  use of algebraic signs, 197
Capillary:
  in bleeders, 154
  as pulsation damper, 149
  for remote indication, 134
Case:
  construction, 28
  definition of, 51
  function of, 108
  materials for, 28, 109
  mounting arrangements provided, 28
  size, 27, 109
Case, cast, 28, 114
Case, drawn, 28, 109, 111, 116
Case, edge reading, 120
Case, flush-mounted (see Case, panel-mounted)
Case, front flange (see Case, panel-mounted)
Case, molded, 28
Case, panel-mounted, 111
  mounting dimensions, 111
  U-clamp, 111, 113
Case, plastic, 28, 120
Case pressure relief, 142-144, 179, 257
  back, 144
  case openings, 143, 251
  grommet, 143
  in high-pressure gas service, 143
  limitations in element rupture, 143, 144
  necessity for, 142
  window, 124, 143

Index   285

Case, sealed, 121, 125
  calculating internal pressure
    in, 126
  pressure relief for, 130
  venting arrangements for, 127
Case, solid back, 117
Case, solid front, 117, 251, 252
Case, turret, 114
Checks, 142, 147
  hole size in, 147
  push, 147
  screw, 147
  use as damper, 147
Cleanliness:
  of compressed air, 185
  in oxygen gauges, 31, 256
  necessity for, 170
  of pressure containing envelope,
    170, 250
Cocks, gauge, 150
Connection, pressure:
  back, 28
  bottom, 28
  definition of, 51
  location of, 28, 52-54, 248
  materials for, 54, 177
  threads, 54, 176, 177, 248
Contacts, electrical, 160
Correction, 197
Correction chart, 195, 198
Corrosion:
  due to atmosphere, 176
  failure due to, 251, 252, 254
  by pressurized medium, 170
Cross ambient effect, 223

Damper, pulsation:
  capillary, 149
  checks, 147
  check valve, 149
  cock, gauge, 150
  in high-pressure systems, 146
  needle valve, 148
  porous metal, 148
  principle of, 146

  purpose, 144
  response time, effect on, 145,
    146
  surge tank, 149
Deadweight tester, 6, 188-193
  accuracy of, 189
    factors affecting, 192
  Bureau of Standards Monograph,
    193
  corrections to, 189
  operation of, 192
  pneumatic, 189
  as pressure source, 185
  reading methods, 197
  theory of, 189
Deadspot, 87
Dial, 52, 97-104
  adjustable, 103
  back-lighted, 103
  cupped, 103
  definition of, 52, 97
  donut, 103
  graduations, frequency of, 102
  graduations, interval of, 173
  materials for, 97
  mirror, 102
  numerals, interval of, 173
  reverse-printed, 103
  with take-up, 98
  with zero band, 101
Diaphragm capsule, 44, 67
Diaphragm plate, 44, 67
Diaphragm, pressure element, 37,
    44, 66-68
  design and fabrication, 66
  linearity of, 46, 66, 71
  materials for, 48, 68
  movement for, 66
  nested, 67
  slack, 71
  stack, 44
  use of, in pressure gauges, 44
Diaphragm seal, 134-142, 209
  cleanout type, 139
  construction of, 134
  damper, use as, 147

Diaphragm seal (cont.)
 diaphragm, protection for, 140
 effect on accuracy, 142, 214
 effect of ambient temperature, 135
 effect of inadequate fill, 134
 fill fluids, 136-137
 filling port, 140
 flushing connection, 140
 maintenance of, 209-214
  filling equipment, 209
  filling procedure, 211-213
  joining seal and gauge, 213, 214
 materials for, 136
 mounting arrangements, 138
 pressure rating, 141
 reason for using, 134
 span limitations, 135
 temperature rating, 141
 volume limitations, 214
Differential pressure:
 definition of, 10
 gauge, 48
DIN standard, 264
Dust, effect of, 176
Dyne, 259

Element, measuring:
 choosing, 71
 definition of, 51
 effect of rupture, 143
 stress in, 249
 types of, 37
Error, 197
 friction, 200
 hysteresis, 201
 linearity, 202
 range, 201
 recording of, 198
 scale shape, 202
 zero shift, 205
Error chart, 195, 198

Failure:
 causes, 184
 corrosive, 251, 252, 254
 element rupture, effect of, 143
 explosive, 170, 251, 252
 fatigue, 144, 250, 254, 256
 overpressure, 250
Failure modes, 250, 254
Fatigue, 44, 60
Fluid, fill:
 compatibility with oxidizing media, 130
 in diaphragm seals, 136
 for liquid-filled gauges, 129, 130
 sanitary, 137
Fluorlube:
 in diaphragm seals, 137
 in liquid-filled gauges, 130, 252
Force:
 relative to pressure, 6
Friction, 200
 correction for, 200
 definition of, 200

Gauge, absolute pressure, 46-48, 67
Gauge, acetylene, 253
Gauge, ammonia, 30, 253
Gauge, chemical, 253
Gauge, clamp-mounted, 28
Gauge, commercial, 29
Gauge, compound, 23
 span of, 32
Gauge, concentric, 19
Gauge, diaphragm:
 application of, 49
 movement for, 66, 93
Gauge, differential pressure, 24, 48, 49
Gauge, dual scale, 24
Gauge, duplex, 23
Gauge, eccentric, 20

Index

Gauge, edge, 120
Gauge factor, 240
Gauge, flush-mounted, 28
Gauge, hydraulic, 31
Gauge, hydrostatic head, 32
Gauge, industrial, 29
Gauge, internally lighted, 120
Gauge, liquid-filled, 125-131, 252, 257
  case pressure, calculation of, 125-127
  case pressure relief for, 130
  case styles, 130
  definition of, 125
  fill fluids for, 129-130
  reasons for use of, 125
  venting arrangements for, 127-129
Gauge, negative pressure, 21
  presentation, 22
  scale for, 98
Gauge, oxygen, 31, 253, 256
  cleanliness requirement, 31, 257
  hazards:
    due to alteration, 31
    due to unknown previous use, 31
  pressure testing of, 185
  pressurizing procedure, 257
Gauge, panel-mounted, 28
  mounting dimensions for, 112
Gauge, pressure, 21
  chemical compatibility, 254, 255
  geometry of layout, 58
  maintenance of, 183
  reuse of, 185, 254
  selection of span, 171
  standard spans, 172
  temperature compatibility, 171, 175
Gauge pressure, 7
Gauge, process, 29
Gauge protectors, 167
  use in test stands, 168
Gauge, receiver, 25

Gauge, refrigeration, 30
Gauge, retard, 24
  span of, 32
Gauge size, 27, 180
Gauge, solid front, 117, 251, 252
Gauge, sprinkler, 32
Gauge, stem-mounted, 28, 248
Gauge, suppressed scale, 25
  span of, 32
Gauge, surface-mounted, 28
Gauge, test, 30, 195
  reading methods, 196
  use as pressure standard, 195
Gauge, vacuum 21, 263
  presentation, 22
  scale for, 98
Gauge, welding, 32
Glycerine:
  in liquid-filled gauges, 129, 252, 258
Graduations, dial, 97
  frequency of, 102
  interval of, 173
Gravimetric units, 14-15, 262
Gravity:
  standard value, 14, 260
  correction for local, 14, 189

Hairspring, 78, 80, 87
  setting of, 200
Hazardous systems, 249, 253
Head correction, 177
  use of bleeder in, 153
Head pressure, 3
  calculation of, 5
  effect on thermometers, 227-228
  units of, 16
Heaters, gauge, 155
Humidity, effects of, 176
Hydrogen embrittlement, 250, 257
Hydrogen, material for use with, 257
Hysteresis, 201
  splitting error, 201

Ideal line, 207, 242
Information supplied to manufacturer, 266
Installation, gauge, 178
International Standards Organization (ISO), 21

Joining, materials for, 75-77

K-Monel, properties of, 61
Kilogram:
  unit of force, 260
  unit of mass, 260
Kilogram per cm² (kg/cm²), 260
Kilopascal (kPa), 15, 260
  conversion to psi, 261
Kilopond, 261

Lazy hand, 157
Leakage:
  degree of, 252
  effect of, 179
Lighting, internal, 120
Linearity:
  of bourdon, 58
  of movement, 58
Link, 80, 88-92
  adjustable (see Link, slotted)
  bimetallic, 90
  bushed, 89
  slip, 89
  slotted, 88
  solid, 88
  temperature compensating, 90, 91, 217
Liquid columns (see Manometers)
LVDT, 235-237
  performance characteristics, 237
  as a pressure transducer, 235-236

Maintenance facility, 184-186
  equipment for, 184
  pressure sources for, 185
  program, 184
  vacuum source for, 185
Manometer, 7, 193-195
  accuracy, factors affecting, 193
  Bureau of Standards Monograph, 193
  mercury, 7, 193
  reading meniscus of, 195
  units of pressure, 193
  use in measuring vacuum, 193
  water, 7, 195
Mass, 15, 260
Materials:
  for bellows, 70
  for bourdon, 61-65
  compatibility of, 255
  for diaphragms, 68
  for diaphragm seals, 136
  for joining:
    nickel alloy, 77
    silver solder, 75
    soft solder, 75
Mechanical shock, effects of, 176
Megapascal, 15, 260
Meniscus, in liquid columns, 195
Mercury:
  density, Table A.5
  manometer, 7
  safe handling of, 194
  in solid-filled thermometers, 219, 221
  limitations on use, 221
Monel, properties of, 61
Mounting, 179
  clamp, 28
  flush, 28
  orientation when, 179, 248
  stem, 28, 109, 179
  surface, 28
Movement, 57, 77-97
  bushed, 80
  components of, 79-80
  damping, 78
  definition of, 52
  diaphragm, 66

Index

Movement (cont.)
  duplex, 93
  gearless, 95
  heavy duty, 81
  lubrication of, 78
  materials for, 80, 81
  plastic, 94
  ratio, 59
  rotary, 81
  vacuum, 86

Negative pressure, 21, 262
  definition of, 9
Newton, 15, 260
Ni Span C, properties of, 64, 65

Ordering information, 266
Overpressure, 60, 81
  effect of, 249
  failure due to, 249, 250
Oxidizing fluids, 170, 252, 258
  hazards of measurement, 170
Oxygen systems:
  hazards in, 25, 252, 253, 254, 256

Pascal, 15
  kilopascal (kPa), 259
  megapascal (MPa), 15, 260
Phosphor bronze, properties of, 61
Pinion, 80, 87
  taper of, 105
Plate, movement, 79
Pointer, 104-108
  adjustable, 105-108
  balance of, 104
  bushing, 104
  definition of, 51
  knife edge, 105
  lifter, 105
  materials for, 104

max-min, 156-160
  construction, 159
  purpose, 156
  retarding effect, 159
Position error, 180, 248
  effect of bulb in thermometers, 227
Potentiometer, 233-235
  as a pressure transducer, 233
  resolution, 234
Pressure:
  absolute:
    converting to, 10-11
    definition of, 8
  atmospheric, 8, 11
    effect of change, 13
    standard, 8, 11, 13
  calculation of, 5
  conversion between units of, 16, 261
  definition of, 3
  differential, definition of, 10
  elements, 37
  gauge:
    definition of, 7
    effect of atmospheric change on, 13
  head, 3-5, 16
  negative, 21, 262-263
    definition of, 9
  operating, 171, 251, 255
  pulsing:
    definition of, 145
    effect of, 144, 255
  steady, definition of, 145
  units of, 8, 14, 260-262
    conversion between, 16, 261
Pressure standards, 186-196
  accuracy of, 187
  line, 187
  measuring, 186
  primary, 187
  reference, 186
  secondary, 187
  for test facility, 186
  transfer, 186

Pressurized medium, 134, 169-173
  chemical compatibility, 170, 250, 252, 254, 255
  corrosive effects of, 170
  oxidizing, 170, 252, 254, 255
  selecting span for, 171
  temperature effects of, 171
Protective shield, use of in test stand, 185
psia, 8
psid, 10
psig, 8

Range:
  adjustment for, 83-85
  error, 201
    correction for, 202
  shift due to temperature, 174
  temperature compensation, 174
Readings:
  alternate method, 196
  application of algebraic signs, 197
  data sheet, 199
  recording of, 198
Regulator, pressure, use in test stands, 185
Resolution, 234-235
  infinite, 235
Response time:
  effect of damping on, 78, 146
  in shut-off valves, 168
Restrictor, 31, 142, 250, 251, 253
Rheostat:
  as a pressure transducer, 230
  resolution of, 234
Ring, 28, 111
  for cast cases, 116
  definition of, 51
  for drawn cases, 113
  friction, 111
  function of, 108
  hinged, 116
  materials for, 111
  for panel-mounted cases, 113
  slip, 111
  snap, 116
  threaded, 111, 121

Scale, dial, 97
  dual, 24
  negative pressure, 98
  vacuum, 98
Scale shape error:
  correction for, 204
  definition of, 202
Sector, 80
  bendable tail, 82, 83
  slotted, tail, 82, 85
Sender, pressure, 229
Sensor, pressure, 229
Shipping temperature:
  effect on thermometers, 227
S.I. Units, 14-15
S.I. System, 14, 15, 259
Silicone:
  in diaphragm seals, 136-137
  in liquid-filled gauges, 129, 252
  use as damper, 78
Silver brazing, 68, 75, 76, 171
Siphon, 150-152
  bulb, 152
  pigtail, 152
  pressure rating, 151
  purpose, 150
Size:
  gauge, selection of, 180
  gauge, standard, 27
Snubber, pressure (see Damper)
Socket, 52-55
  bar stock, 54
  cast, 55
  center-back, 53
  definition of, 51
  forged, 55
  low, 53
  low-back 53

Index

Soft solder, 68, 75
  temperature limitation, 75, 171
Span:
  of compound gauge, 32
  definition of, 32
  preferred, 171
  of retard gauge, 32
  selection of, 171
  of suppressed gauge, 32
Stainless steel, properties of, 64
  cases, 109, 116, 119
  joining, 76
Standard for pressure gauges:
  B40.1, 20, 34, 247-254
  DIN, 264
  foreign, 263
Start (see take-up)
Steam tracing, 156
Stops, 80, 89, 92
Strain gauge, 238-240
  basic circuit, 238
  diffused silicon, 239
  performance characteristics, 239
  as pressure transducer, 238
Surge tank, as pulsation damper, 149
Switches, electrical, 160, 167
  abutting contacts, 160
  applications of, 160
  arcing of, 163
  circuits, 163
  differential, 163, 167
  explosive hazard, 163
  high-low limit, 164
  magnetic reed, 165
  normally closed, 163
  normally open, 162
  retard effect, 160, 161, 163
  snap action, 161, 163
Système International d'Unites, le, 15, 259

Take-up, 98-101, 256
  reading scales with, 100

Tapping, 180, 196
Technical atmosphere, 262
Temperature:
  compensation for, 174
  effects of, 174
  limits, 175
Testing:
  definition of, 184
Thermometer, gas-filled, 215-219
  ambient pressure, effect of, 227
  ambient temperature, effect of, 227
  bulb volume, effect of, 216
  filling medium for, 215
  head temperature compensation, 217
  limitation of span, 219
  shipping temperature, effect of, 227
  theory, 215
Thermometer, liquid-filled, 219, 221
  ambient pressure, effect of, 227
  ambient temperature, effect of, 220
  bulb position, effect of, 228
  bulb volume, 219
  filling media, 219-221
  filling pressure, 221
  head temperature compensation, 220
  shipping temperature, effect of, 227
  theory, 219
Thermometer, vapor pressure, 221, 226
  ambient pressure, effect of, 227
  ambient temperature, effect of, 226
  bulb position, effect of, 227
  bulb volume, 225, 226
  cross ambient effect, 223
  filling media, 222-223

Thermometer, vapor pressure (cont.)
  head temperature compensation, 226
  nonlinearity of, 222-223
  shipping temperature, effect of, 227
  theory, 221
Tip, 71-73
  definition, 51, 71
  design, 72
  fabrication, 73
  joining, 73
Tip travel, 40, 56-57, 59-60
  adjustment for, 83
Torr, 262
Torricellian vacuum, 13
Transducer, capacitance, 240-241
Transducer, pressure, 196, 229-246
  accuracy, 242
  ambient temperature, effect of, 244
  automotive type, 230-233
  definition of, 229
  performance specification, 245
  specifying accuracy, 242
Transfer function, 230
Transmitter, pressure, 229

U-clamp, 111, 113, 114
Units of pressure:
  conversion between, 16, 148, 260, 261

Vacuum:
  definition of, 9
  movement, 86
  source, 185
Valve, check, as pulsation damper, 149
Valve, needle, as pulsation damper, 148
Valve, shut off, 167
Vapor pressure:
  ethyl ether, 222-223
  nature of, 221
  propane, 222-223
  thermometers, 221-226
Venting:
  case (see Case Pressure relief)
  prior to opening containers, 256
Vibration, effect of, 176, 255

Water:
  density of, Table A.5
Welding, 76
Wheatstone bridge, 238
Window, 122
  beveled, 123
  crowned, 124
  definition of, 51, 122
  flat, 122
  hazard of, 251, 253
  one-piece ring and, 123
  plastic, 124
  pressure relief, 124

Zero band, 101
  reading scales with, 102
Zero shift:
  compensation for, 175
  due to temperature, 174
  error, definition, 205
  correction for, 206